THE LITTLE BOOK *of* EARTHQUAKES *and* VOLCANOES

THE LITTLE BOOK *of* EARTHQUAKES *and* VOLCANOES

ROLF SCHICK

COPERNICUS BOOKS
An Imprint of Springer-Verlag

Originally published as *Erdbeben und Vulkane*,

Published in the United States by Copernicus Books,
an imprint of Springer-Verlag New York, Inc.
A member of BertelsmannSpringer Science+Business Media GmbH

Copernicus Books
37 East 7th Street
New York, NY 10003
www.copernicusbooks.com

Library of Congress Cataloging-in-Publication Data
Schick, Rolf.
[Erdbeben und Vulkane. English]
The little book of earthquakes and volcanoes / Rolf Schick.
p. cm.
Includes bibliographical references and index.
ISBN 0-387-95287-X (h/c : alk. paper)
1. Earthquakes. 2. Volcanoes. I. Title.
QE521.S35 2002
551.22—dc21 2002023144

Manufactured in the United States of America.
Printed on acid-free paper.
Translated by Peggy Hellweg.

9 8 7 6 5 4 3 2 1

ISBN 0-387-95287-X SPIN 10838251

Preface

At the end of a trip by airplane or ship, we often exclaim, "Finally, back on solid ground!" But how solid is the Earth on which we stand, really? The imprints of seashells in the rocks of mountains and the fossil remains of tropical animals in polar regions suggest that land masses do move. From the perspective of a human lifetime, however, the motion is so slow as to be imperceptible. But occasionally the solid surface changes suddenly. Then earthquakes pull the apparently stable ground from under our feet, or volcanic eruptions show us that the surface on which we live is really just the thin skin of the Earth. When these events occur, we speak of natural disasters. But is this term truly justified? People are not killed directly by the ground's movement, but by the collapse of man-made structures. And eruptions contribute more to maintaining life than to destroying it. Volcanism replenishes the water in the oceans, and the most fertile soils are found in volcanic regions.

The movement and deformation of the solid Earth are products of its inner heat and the resulting temperature distribution. Although the heat flow to the surface from the Earth's interior is only 1/18,000 of the heat from the Sun, it is just as important a prerequisite for human life. A thermal engine is working under our feet. Without this engine's mountain-building forces, the Earth's surface would completely disappear under the oceans within a few tens of millions of years. No volcanoes would maintain the supply of carbon dioxide, which helps to moderate our world's climate, and which is vital to the process of photosynthesis. In about 500 million years, long before the Sun's energy begins to fade, the Earth's inner temperature will drop below a specific limit, causing the processes we call *tectonics* to stop. Conditions for life on Earth will change drastically.

This book describes the intermittent but rapid tectonic processes that lead to earthquakes and volcanism. These processes are associated with complex, irreversible dynamics, which make a quantitative description difficult. The causes of earthquakes and volcanic eruptions lie hidden from direct observation, deep within the Earth, and can only be investigated using indirect procedures. In addition, each earthquake and each volcanic eruption is a unique, irreproducible process whose occurrence cannot be predicted.

Under such circumstances, it is not surprising that we can describe earthquakes and volcanism only approximately in a mathematical and physical sense, and that we must use idealized models to try to understand them. Nonetheless, earth scientists remain fascinated by the glimpses of order and regularity that emerge from the chaotic dynamics of these processes.

Let us, in Chapter 1 of this book, explore the structure and heat budget of the Earth and the resulting dynamic processes. In Chapters 2 and 3, we will then examine the historical development of our understanding of earthquakes and volcanoes and discuss their fundamental causes and effects as well as the possibility of predicting their occurrence. And at the end of the book, you will find a list of readings and Internet resources for further exploration of the subject.

Rolf Schick
Stuttgart, Germany, February 2002

Acknowledgments

I wish to thank Hans Berckhemer (Frankfurt), Götz Schneider (Stuttgart), and the late Stephan Müller (Zurich), all of whom aroused my interest and enthusiasm for earthquake seismology. I also want to thank Marcello Riuscetti (Catania and Udine), who convinced me that—even for a physicist—volcanoes can be more than "smoke and fire." I appreciate the help of my son Peter, who assisted me in the preparation of the figures. The greatest acknowledgement goes to my wife Inge for her continued encouragement and understanding.

chapter 1

How Solid Is the "Solid" Earth?

However still the Earth's surface may seem at times, it is actually seething with activity, much of it driven by the intense heat of the inner layers of the Earth. In this chapter, we will learn how the Earth came to have this heat and why it is now cooling. We will also discuss how the changing temperature affects the movements of the plates that make up the Earth's crust and how, over the years, scientists have developed a greater understanding of all of these phenomena.

The Earth's Temperature

On a train from Zurich to Milan you will pass through Switzerland's Gotthard Tunnel, a tunnel 15 kilometers long and up to 1000 meters below the surface through a granite massif. If the day is cold and wintry, you may notice that shortly after you enter the tunnel, the air coming in through your window is no longer frigid, but warm, almost subtropical. Surprisingly, we did not become aware of this continuous, worldwide flow of heat rising from deep within the Earth until the beginning of the nineteenth century. Up until then, hot springs and glowing lava from volcanoes were thought to be local phenomena related to fires burning just beneath the Earth's surface. This belief only changed around 1830 with the discovery of a 1 °C-rise in temperature for each 30 meters of descent into a deep mine for metal ore. This mine was located in the German state of Saxony, far from volcanoes and underground coal seams, where burning coal could have caused a rise in temperature, as was generally assumed at the time. Similar increases in temperature were also measured shortly afterward in French and English mines. Even by today's standards, these measurements represent surprisingly precise determinations of the rise in the temperature with increasing depth below the Earth's surface in continental areas.

What determines the temperature distribution in the Earth? Even today, many theories about the thermal and material development of our planet are hotly debated. Most earth scientists, however, agree with the following picture. The Earth, like the other planets of the Solar System, grew from the rapid accumulation of *planetesimals.* These collections of cosmic debris, each 10 to 100 kilometers in size, are similar to the asteroids in the belt between Mars and Jupiter. With each collision between a planetesimal and the Earth, kinetic energy was converted to heat, and the Earth's temperature increased as it grew in volume. When the temperature was high enough, the Earth's interior melted. Within a relatively short time, perhaps 100 million years, gravity caused the components making up the molten Earth to separate according to their densities. The iron-rich, heavy core that formed was surrounded by the lighter silicates of the Earth's mantle. During this epoch, the Earth contained a high level of thermal energy and sustained high temperatures. The Earth has been cooling ever since.

Let us consider the factors responsible for the heat loss. A planet loses thermal energy to space by radiation from its surface. The loss of heat depends primarily on the ratio of the surface area of the planet, A, to the mass, m, which can hold the heat. The larger the value of A/m, the more rapidly the planet loses heat. If we set the ratio of A/m equal to 1 for Earth, then the ratio is 6 for the Moon, 2.5 for Mars, and 1.1 for Venus. This is the main reason why tectonic movements on the Moon came to an end more than 1 billion years ago, and why tectonics on Mars has probably not been active for more than 100 million years.

Tectonics is what we call the deformation and movement of material caused by forces acting inside a planet. Without tectonics, there can be no earthquakes or volcanic activity. In contrast to the Moon or Mars, tectonics has been active on Earth for the past 1 billion years. Due to the rise in temperature with increasing depth, our planet's interior has developed a hot, plastic *asthenosphere* that flows underneath the more or less rigid cover of a cold *lithosphere*. The material in the asthenosphere moves by convection and transports heat from inside the Earth to its surface.

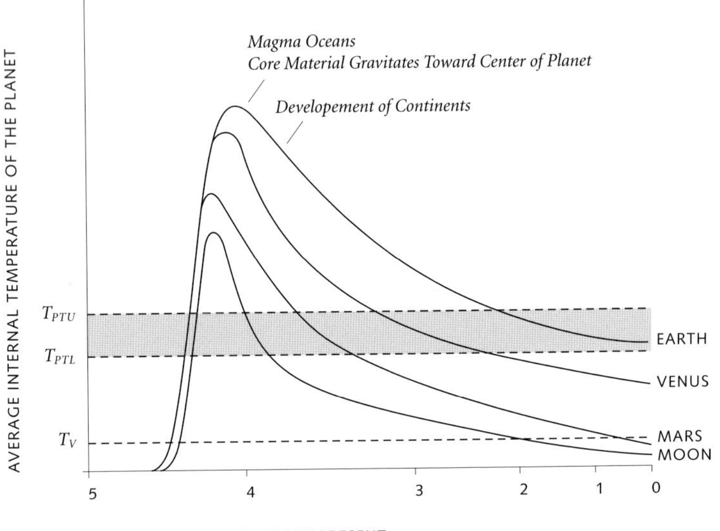

The thermal development of Earth-like planets.

The chart shows the development of the mean temperature of the terrestrial (Earth-like) planets and the Moon from their creation to the present. The curve for Mercury is nearly the same as that for the Moon. T_{PTU} and T_{PTL} give the upper and lower bounds for the temperatures between which the convection occurs that drives plate tectonics. T_V is the minimum temperature for the existence of volcanism.

(After a diagram by K. C. Condie.)

The Earth's *mantle*, which includes the asthenosphere, is the thick layer of material between the crust and the core. It acts like a giant heat engine in which the movement of heat and material from deeper, hotter rocks to the Earth's cooler surface is accompanied by mechanical forces. The spectacular effects of these forces are earthquakes and volcanoes. Less obvious to us, but just as important, are the almost unnoticeably slow processes of the movement of continents, the creation of new ocean floor, and the rise of mountain ranges. All these processes are caused by convection. They are explained by the theory of *plate tectonics.* We know from daily life that convection, the transfer of heat accompanying the movement of substances, is much more efficient than heat conduction. Most of the heat that reaches the Earth's surface and is radiated into space is transported from its interior by convection processes in the mantle.

Not all the heat that is lost was generated during the creation of the Earth. Some heat is still produced today by the decay of radioactive isotopes in the Earth's crust and mantle. Most of the heat produced by radioactivity comes from the decay of isotopes with long half-lives, such as uranium-238, thorium-232, potassium-40, and uranium-235. These radioactive elements are 100 times more common in the upper crust than in the mantle. Still, the amount of heat produced by radioactive decay in the mantle is more important because the mantle has a much larger volume than the crust.

Probably there are other sources of heat inside our planet. Some scientists believe that the Earth's liquid outer core, which consists primarily of iron and nickel, is more or less free of radioactive decay products. However, in order to maintain the dynamo that causes the Earth's magnetic field, rapidly moving convection currents must exist. The necessary temperature gradient is apparently caused by the release of latent heat during the "freezing," or solidification, of liquid iron at the interface to the Earth's solid inner core.

Recent detailed measurements of the Earth's heat flow, which were made worldwide, have allowed us to improve our estimate of the heat loss of the entire planet. The present estimate for the average heat flow, 80 milliwatts per square meter, is nearly twice as high as the estimate made in 1970. At this rate, the heat lost globally is clearly higher than the amount of heat currently produced by radioactive decay. This loss of heat does not necessarily imply that the temperature of the Earth's interior is decreasing, since there is still plenty of heat in the molten outer core. In the long run, however, the Earth is cooling. According to the latest estimates, in about 500 million years our planet will no longer be able to support tectonic activity.

The Structure of the Earth and the Role of Plate Tectonics

Our planet's interior is dominated by changes that occur as the depth beneath the surface increases. Rocks of different chemical composition or different physical state meet along

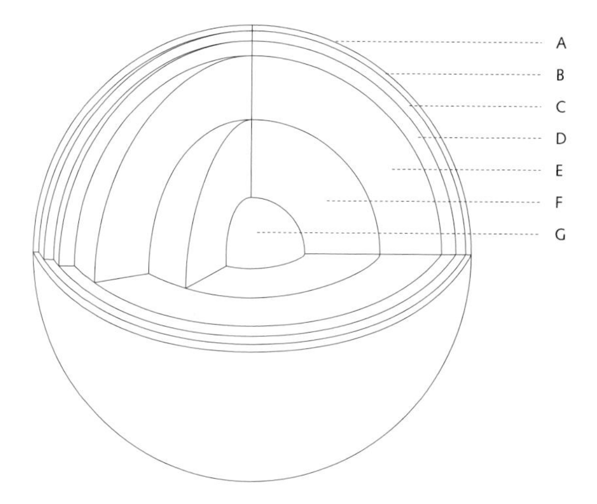

The layered structure of the Earth.

A. Crust. Depth range: 6 kilometers under the oceans (mainly basalt), 30 to 60 kilometers under the continents (upper crust: granites and gneiss; lower crust: basalt). The Mohorovičić discontinuity separates the Earth's crust from the mantle, which consists mainly of ultramafic rocks such as olivine, pyroxene, and garnet.

B. Subcrustal lithosphere. Depth range: 0–80 kilometers to 150 kilometers.

C. Asthenosphere. Depth range: 80 kilometers to 150–400 kilometers.

D. Transition zone where rocks under high pressure change their crystal structure. Depth range: 400 kilometers to 660 kilometers.

E. Lower mantle. Depth range: 660 kilometers to 2900 kilometers.

F. Outer core. Depth range: 2900 kilometers to 5150 kilometers. Consists mainly of iron, some nickel, oxygen, and/or sulfur. Very low viscosity, "liquid."

G. Inner core. Depth range 5150 kilometers to 6371 kilometers. Consists mainly of iron and some nickel. High viscosity, solid.

(After a drawing by K. Strobach.)

horizontal or slightly dipping interfaces. Primarily, our knowledge of the Earth's inner structure has been derived from investigations of the propagation of seismic waves, which are generated by earthquakes or artificially induced by explosions. The first to determine that there are seismic boundaries, or *discontinuities*, which divide the Earth's interior into different layers, was the Yugoslavian seismologist Andrija Mohorovičić. The discontinuity he discovered in 1909 separates the lighter rocks of the crust from the denser material of the Earth's mantle below it. This boundary is now called the *Moho*. The depth of the Moho depends on whether it is located under high mountains, under a plain, or under an ocean. Under the Rocky Mountains, for example, the Earth's crust is about 60 kilometers thick, while it is only about 40 kilometers thick under the Great Plains, and 6 kilometers under most of the Atlantic Ocean. In 1911, Beno Gutenberg, a German seismologist who later moved to the U.S., discovered another discontinuity at a depth of 2900 kilometers. This transition zone, now called the *D″-layer*, separates the solid silicate rocks of the mantle from the Earth's core, which mainly consists of iron and nickel. In 1936, the Danish seismologist Inge Lehmann was able to prove that the core is divided into a liquid outer and a solid inner part. The boundary between these two layers is now called the *Lehmann discontinuity*.

The discovery of our planet's layered structure, however, cannot explain the variety of landforms we actually *see*. Why is the Earth's surface with its plains, mountain ranges, island arcs, ocean basins, and trenches so heterogeneous? Why does it seem to deviate from the symmetry of the Earth's sphere? And why are earthquakes and volcanoes not spread evenly over the Earth's surface?

The discovery of fossil shells in the high Alps was a sign to early geologists that the rocks of which they are formed must have risen from under the sea. On the other hand, massive layers of debris caused by erosion, sometimes up to 10 kilometers thick (e. g., in the Po River Plains in Italy), indicate that other areas are now lower than they were ages ago. These two examples give evidence of strong vertical movements of the Earth, with some places rising while others subside. These observations were initially taken as evidence for the hypothesis that the radius of the Earth was shrinking as our planet cooled. In 1912, the German meteorologist Alfred Wegener theorized that the continents were wandering over the surface of the Earth. This caused heated discussion among his fellow scientists. Eventually, contemporary researchers rejected Wegener's theory because they believed that the mantle had to be very rigid in order to support mountain ranges, such as the Himalayas or the Alps, over millions of years. The assumption of a rigid mantle precluded Wegener's hypothesis of a horizontal movement of entire continents, since the forces that would be necessary to move them were unimaginably large. This assumption was accepted for several decades, during which no significant scientific progress was made.

The end of World War II heralded the beginning of intensive investigation of the oceans and the seafloor. The Mid-Atlantic Ridge was found to be a mountain range of basalt stretching from the Arctic to the Antarctic. In only a few places do its peaks rise above the water. The Pacific and Indian Oceans had similar submarine ridges. In the Pacific, the American marine geologist Harry Hess found low-lying, truncated volcanic cones, which he named *guyots* after the first professor of geology at Princeton University, Arnold Henri Guyot. Hess recognized that the guyots are ancient, more than 500-million-year-old volcanoes, which were carried over thousands of kilometers away from the place they were formed by a movement of the seafloor. But many leading earth scientists refused to accept a relationship between his observations and a movement of the crust, which Hess associated with the rise of basaltic melt (magma) from the mantle through cracks in the thin oceanic crust along the ridges. The discussion finally came to an end in 1964, when the British geophysicists Fred Vine and Drummond Matthews measured alternating polarities in the rocks along the mid-ocean ridges. Since these magnetic polarities fit with the known reversals of the Earth's magnetic field, Vine and Matthews were able to confirm the hypothesis that basaltic melt rises through the ridges and forms new oceanic crust, which moves to both sides at a rate of several centimeters per year and takes on the current magnetic polarity. This new theory of *seafloor spreading* was confirmed through subsequent investigations.

But doesn't the continuous rise of magma from the mantle mean that the surface of the Earth is growing larger? The possibility of an expanding Earth had often been discussed in the earth sciences, and the concept of seafloor spreading briefly encouraged the proponents of this theory. Before long, though, a significant mid-twentieth-century discovery was recalled to complement the idea of seafloor spreading. Back then, seismologists Kiyoo Wadati and Hugo Benioff from Japan and the U.S., respectively, had discovered regions at continental margins and along island arcs where unusually large numbers of earthquakes occur along narrow zones that penetrate deep into the mantle. These earthquakes could mark the path of crustal plates sinking into the Earth.

The recognition of both the spreading of the ocean floors and the sinking of crustal plates opened the way for the development of a concept that explains all large-scale processes in the Earth's outer layers—the model of *plate tectonics.* One fundamental discovery in the field of plate tectonics is that of a broad transition zone located at a depth of about 100 kilometers, where both the temperature gradient and the viscosity of the rock change. This boundary marks the transition from the cool and brittle outer layer called the *lithosphere* to the hot, plastic, and viscous *asthenosphere,* a layer that is subjected to long-lasting forces. The lithosphere is divided by the Moho into a crustal and a subcrustal part, so that the subcrustal lithosphere is part of the Earth's upper mantle. The underlying asthenosphere reaches a maximum depth of 660 kilometers.

If the lithosphere is covered by continental crust, we call it continental lithosphere; otherwise, oceanic lithosphere. The entire lithosphere is divided into individual lithospheric *plates.* On the whole, these plates are mechanically rigid, but can be moved by the currents of the asthenosphere. Since the transition zone that acts as a lubricant between the lithosphere and the asthenosphere is located at a greater depth than the Moho, which, in turn, lies deeper underneath a continent than underneath an ocean, the plate boundaries do not always coincide with the edges of the continents. While the centers of the plates are considered to be rigid, the plate boundaries may be deformed. At first glance, this assumption does not appear to be very meaningful, but it was the key to success for the model of plate tectonics: Since all tectonic interactions are limited to the edges of the plates, it is at the plate boundaries where tectonic processes (i. e., the formation of mountains and rifts as well as earthquakes and volcanism) are bound to occur.

The theory of plate tectonics was completed by the solution to a simple problem of spherical geometry. In the eighteenth century, the mathematician Leonhard Euler investigated the movement of sections on the surface of a sphere. The movement of each section can be described relative to an axis passing through the center of the sphere. The point where the axis meets the surface of the sphere is the pole of rotation for that section. In 1968, the Englishman Dan McKenzie and the American Jason Morgan each independently applied this concept to the spherical Earth and derived three fundamental modes of movement for the edges of the lithospheric plates:

1. The lithospheric plates diverge—that is, they are separating. Hot, fluid material from the mantle is pressed into an opening crack and welded to the edges of the

plate. A mid-oceanic ridge develops. The newly formed oceanic crust drifts to the sides, typically at a velocity of 2–3 centimeters per year. This is the process known as seafloor spreading. Due to the spherical geometry of the separating plates, the crack may be interrupted by offsets, called *transforms*. The Mid-Atlantic Ridge and the East Pacific Rise are examples of this process of divergence. At the East African Rift and the Gulf of California, divergence is just beginning.

2. The plates converge—that is, they are moving toward each other. In a collision, the lighter plate, usually the continental plate, pushes itself onto the heavier oceanic plate. The three possible interactions depend on the types of plates involved: oceanic–oceanic, oceanic–continental, or continental–continental. In the first case, island arcs such as Indonesia or Japan are created. Under the Andes of South America, the oceanic Pacific plate is being pressed under the continental plate, a process called *subduction*, and the long mountain range of the Andes is developing along the edge of the continent. When two continental plates collide, there is no subduction. Rather, compressive forces in the nearly equally heavy and thick continental plates produce great mountain ranges. The Himalayas are an example of this type of collision.
3. The plates slide horizontally past each other. The results are vertically oriented, often very long, long-lasting, and usually very active earthquake faults. They are called *transform faults*, a term created by the

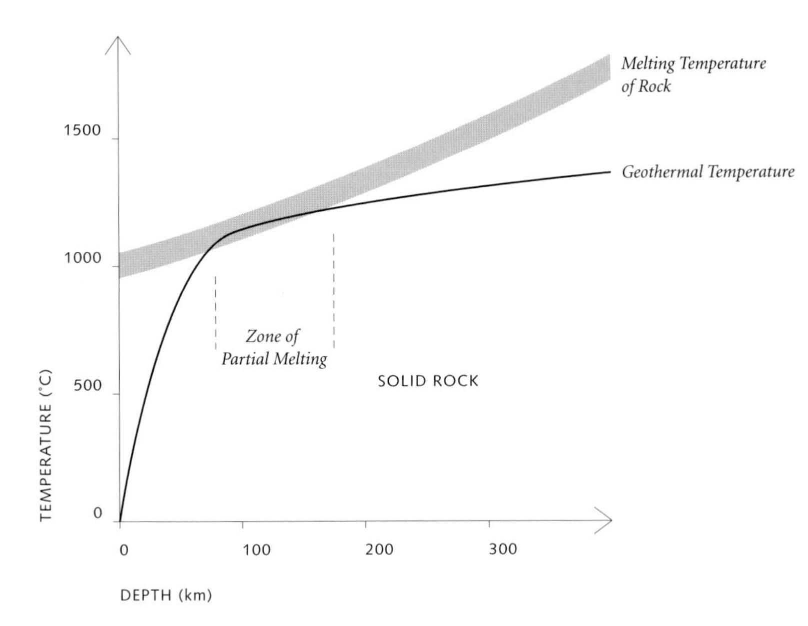

Geothermal temperature and average temperatures at which rocks begin to melt within the upper 400 kilometers of the Earth's body. For depths greater than 100 kilometers, the temperature increases more slowly with increasing depth. This marks the transition zone between lithosphere and asthenosphere. The decrease in the temperature gradient in the asthenosphere is caused by the "stirring" of convection that begins at this depth. The melting temperature of rocks increases with pressure. At the transition from the lithosphere to the asthenosphere, the geothermal temperature reaches the melting temperature of the rock matrix. The components of the matrix with the lowest melting temperatures melt. The resulting *partial melts* are the source of volcanism.

> Canadian geophysicist Tuzo Wilson. Transform faults often connect the movement between the offset segments of divergent plate boundaries. Most of these faults are found on the seafloor. The well-known San Andreas Fault is a transform fault.

The concept of plate tectonics brought satisfying order to our understanding of the large-scale structures of our planet. One of its successes was the confirmation of the plate velocities that the theory had predicted. Highly precise, continent-spanning geodetic methods, such as those using the Global Positioning System (GPS), have determined that the plates are moving at velocities of 1 to 10 centimeters per year—just as predicted.

chapter 2

Earthquakes

For thousands of years, most people believed that earthquakes were mythical or spiritual occurrences, triggered by unseen powers. Over the past hundred years, scientists have been able to explain these complex events in more concrete terms. In this chapter, we will chart the history of the study of earthquakes and explain the methods and analytical tools scientists are using today. Then we will discuss the challenges of predicting earthquakes and designing structures that are capable of withstanding their power.

Early Ideas and the Concept of "Tectonic Quakes"

In the myths of many societies, earthquakes were attributed to animals living under the ground. The low, rolling noises, the sudden shifting of the ground followed by wave-like movements of the Earth's surface, the cracks left behind—all these phenomena suggested that their source was a monster in the depths of the Earth. The Japanese thought it was a scorpion. People in India blamed a salamander, and North American Indians, a turtle. The Maori of New Zealand explained earthquakes as the kicking of an unborn child in the body of Mother Earth. In Greek mythology, Poseidon, the god of the sea, was called the "Earth shaker" or "Earth holder," probably because the Greeks thought the Earth was a disk floating in water. The waves generated by undersea earthquakes in the Mediterranean Sea often caused extensive damage along the coasts and probably encouraged this view of their source.

Based on their observations of Nature, Greek and Roman philosophers and scientists tried to find an explanation for the Earth's movements. Some of their views were not unreasonable. Anaxagoras (500–428 B.C.) thought that earthquakes were caused by the shell of the Earth collapsing, partly due to washouts and partly because mountains were hollowed out by underground fires. According to Aristotle (384–322 B.C.), earthquakes were caused by air caught in underground caves. When the air tried to escape from the caves, the Earth shook.

As with many of Aristotle's ideas, most Europeans believed his explanation of earthquakes up until the end of the Middle Ages. There were, however, some who fanatically opposed the ideas of Aristotle and proposed superstitious or fantastic explanations for the Earth's movement or took it to be a sign of the Apocalypse. For example, after a strong earthquake near the Lower Rhine in 1682, the Flemish alchemist Jan Baptista van Helmont opined that the wings of an avenging angel beating the air produced a sound that made the Earth shake.

After the discovery of electricity, earthquakes were often assigned electrical causes. People suggested that pyramid-shaped buildings could act as lightning rods for underground thunderstorms and thereby protect against ground-shaking.

At the beginning of the nineteenth century, the German naturalist Alexander von Humboldt made extended journeys through volcanic areas. He noted that volcanic eruptions seem to be accompanied by many, mainly weak earthquakes, while large earthquakes mostly occur in regions with no volcanoes. In an extension of Aristotle's idea, von Humboldt formulated a theory that remained popular well into the twentieth century even though it was incorrect. He thought earthquakes were caused by forces from expanding underground volcanic gases. Von Humboldt wrote, "One could say, the Earth shakes more strongly when there are fewer airholes in the surface of the ground."[1] He saw volcanoes as safety vents for releasing the volcanic forces spread throughout the Earth. The respect accorded to von Humboldt encouraged the rapid spread of his idea. Following von Humbolt, members of the "Plutonist school" of thought believed that earthquakes and volcanoes were different parts of a single process.

At first there was little disagreement with these concepts. One of the most prominent critics of the Plutonists was Georg Heinrich Otto Volger, a professor at the University of Frankfurt who studied earthquakes in Switzerland. Volger did not deny the existence of volcanic quakes. However, he believed that earthquakes outside of volcanic areas were generated by the collapse of underground caverns caused by the leaching of ground water. His thoughts appeared in the literature as the *hollow layer hypothesis.* With these opinions, Volger joined the "Neptunist school," a group of geologists who thought that sedimentation and chemical deposition from water were the main causes of changes on the face of the Earth. Plutonists and Neptunists fought each other's ideas, both sides refusing to make even the slightest concession.

After 1865, a group of Austrian alpine geologists and earthquake researchers under the leadership of the Viennese geologist Eduard Suess worked out a completely different concept for explaining the cause and mechanism of earthquakes. Studying earthquakes in the Alps, they recognized a connection between the affected areas and geological fault lines, which are fractures in the crust along which rock is displaced. When Suess determined the locations of a sequence of earthquakes in Calabria (southern Italy), he found a systematic association of their locations with rock displacements. The dislocations lay along geological faults where layers originally deposited as a single unit were offset by the displacement along the fault. For the first time, groups of earthquakes and their effects were related to the geological and tectonic structure of the region in which they occurred.

Rudolf Hoernes, a professor at the University of Graz, Austria, and a nephew of Eduard Suess, studied earthquakes in the Near East. In 1893, he wrote:

> There can be no more doubt that we must view the seismic events occurring so frequently and with such appalling effects in Palestine and Syria as tectonic earthquakes that are associated with the immense dislocation [of the Jordan Rift Valley] . . . and it is clear that the earthquakes of the Jordan region cannot be the result of collapses. . . . We recognize the intimate relationship between movements in the rock matrix and the shaking, which we call tectonic or dislocation quakes.[2]

From his observations, Hoernes concluded that the quakes associated with collapses and volcanoes are smaller and much less ubiquitous than the earthquakes he called tectonic. We still use the classifications that he developed: collapse quakes, volcanic quakes, and tectonic quakes. Hoernes included the first two categories to avoid attacks from their supporters.

Such caution was probably not necessary. Within a short time, there were many reports of displaced rocks at earthquake-caused cracks. Over a distance of several kilometers in the Neo Valley in Japan, an 1891 earthquake shifted dams between rice fields more than 4 meters horizontally. The great earthquake of 1906 in San Francisco brought final proof of the tectonic nature of earthquakes. By chance, extensive cartographic work had been done in the area around San Francisco shortly before the quake. By comparing before-and-after measurements, Harry Fielding Reid, a professor of geology at the Johns Hopkins University, developed his *theory of elastic rebound.* This concept, which is still accepted today, explains the kinematics and mechanics of what happens when an earthquake occurs.

The Mechanics of Earthquakes

Let us consider Reid's earthquake model in greater detail using the San Andreas Fault near Hollister, California. Hollister is a particularly good example because the fault is narrow and its rate of movement is high. Imagine that rows of trees are planted across the fault. After only a few years, we can see that the rows of trees are displaced where they cross the vertical fault.

The San Andreas Fault runs from northwest to southeast and is the junction between the Pacific plate to the west and the North American plate to the east. The rate of displacement of the plates relative to each other is about 5 centimeters per year. If the plates can move relatively freely past each other, no large earthquakes occur. If the plates are stuck together, then stress builds up in the region on both sides of the fault due to the continuing relative motion of the plates. Similar to the stretching of a spring, the kinetic energy of the plate motion is stored as potential energy in the deformation of rocks. The material at the fault is subjected to the maximum shear stress. The stress must be absorbed by the strength of the material between the crustal blocks that form the two sides of the fault. The strength is primarily determined by adhesion and friction between them. If the shear, or tangential, stress exceeds the strength of the material, then the material breaks at the fault. We call this a *shear rupture* because it is caused by shear stresses. The new rupture does not necessarily have to follow the old fault. However, compared to the surrounding material, the material along the existing fault is more fractured and therefore weaker. Thus, new ruptures tend to follow previous ruptures, concentrating the fault in a narrow zone.

During the beginning of an earthquake, the friction between the blocks is drastically reduced and they can move almost freely along the crack. Like a steel spring released from tension, the crustal blocks jump to a new equilibrium. In just a few seconds or minutes, Nature suddenly creates a displacement that would have happened gradually in tens or hundreds of years if the crustal blocks could always move freely past each other. The edges of the blocks again become stable in the new equilibrium position. They achieve a new strength, and the process of loading the system with deformation energy begins again. The forces affecting the rock matrix *before and after* the rupture change so slowly that we consider them to be static. *During* the rupture, they are converted into sudden accelerations, or dynamic forces. These forces are no longer related to static, but rather kinetic energy.

This essentially reflects Reid's theory of elastic rebound. But Reid could not find an explanation for the source of the necessary forces. At the time, this was not a big problem because the idea that the Earth was cooling and shrinking left open the possibility for cracks that could lead to earthquakes.

In the light of modern plate tectonics, we can state that the Earth is a thermodynamic machine, which converts heat energy into mechanical forces that causes earthquakes. The cold and brittle material of the lithosphere rides on the convection currents in the Earth's upper mantle. The mantle's movement leads to the storage of potential energy in the lithosphere in the form of slowly increasing deformation. A shear rupture in the rocks transforms the potential energy into a pulse of kinetic energy. Part of this energy leads to the shaking of the Earth's surface—earthquakes. The process of storing and releasing energy often occurs in earthquake cycles. Typically, the energy for a large earthquake accumulates over 100 years or so and is discharged in a time frame on the order of 1 minute.

Such cycles of transfer of energy from one form to another are common in Nature. Lightning, hurricanes, landslides, avalanches, and volcanic eruptions are all the results of similar sequences. The sudden conversion of energy is based on an instability in the system. For a lightning strike, it is the electrical resistance of the atmosphere; for earthquakes, it is the shear strength of the rocks; for volcanic eruptions, it may be an instability in the flow of magma.

The investigations of the great San Francisco earthquake of 1906 by Reid and his colleagues can be considered the birth of the modern explanation of earthquake sources. At the time, most geologists and seismologists viewed his results with suspicion. This is still apparent in the name of Reid's *shear rupture hypothesis.* Can the rapidly occurring dislocations really be the source of earthquakes and not

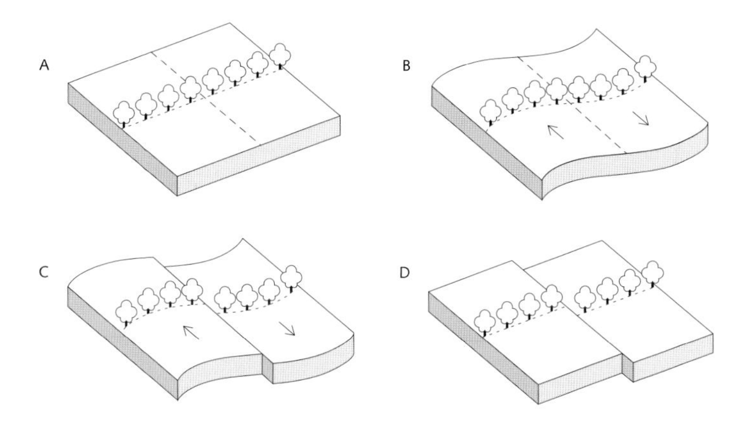

An earthquake is the result of a sudden dislocation of mechanically stressed crustal blocks.

A. Initial situation: Two crustal blocks border each other along a fault (the dashed line perpendicular to the row of trees). The straight row of trees shows the unstressed condition at the beginning of the cycle.

B. Increasing tension: On both sides of the fault, the blocks are deformed by the shear forces. The strongest forces occur along the fault.

C. Shear faulting: The tension exceeds the shear strength of the rocks along the fault and it ruptures. There is a sudden dislocation along the fault line.

D. Final situation: Through elastic rebound, a new equilibrium develops in the crustal blocks. It is equivalent to the initial situation.

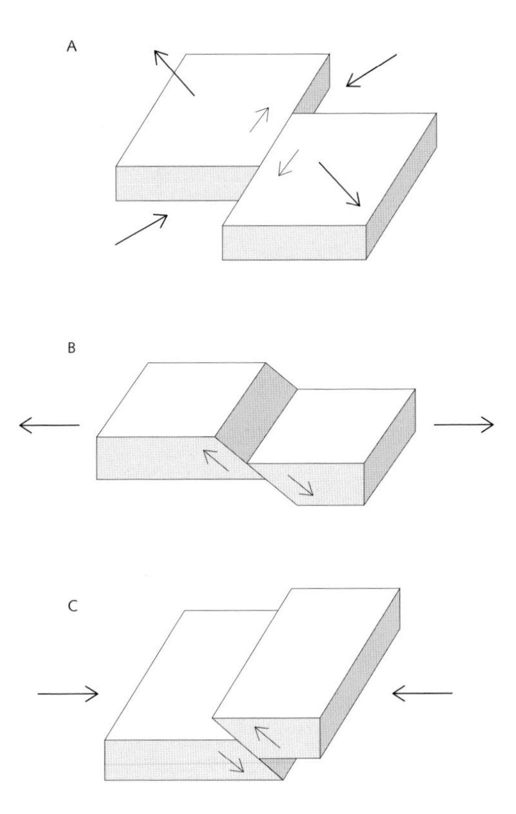

Basic types of displacement in earthquakes.

The position of the shear plane in the earthquake source depends on the direction of the tectonic stresses in the source region. These stresses are described by the stress tensor.

A. Strike-slip fault. Compression and extension are both oriented horizontally, which leads to the development of a horizontal dislocation on a vertically oriented shear plane. The earthquakes along the San Andreas Fault are predominantly of this type.

B. Normal fault. Extension is horizontal, while compression is oriented vertically. Earthquakes of this type are found in areas with rifting. The horizontal extension promotes the rise of magma.

C. Thrust fault. Compression is horizontal, while extension is oriented vertically. Earthquakes of this type are found where mountains are forming and growing.

simply the reflection of some unknown source of energy deep within the Earth? Are the many small earthquakes that occur without producing any signs at the Earth's surface generated by the same mechanism? These discussions were renewed in 1922, when the English seismologist Herbert Hall Turner located earthquake sources at depths of 100 kilometers and more. At these depths, the pressure must force the rocks together so strongly that it would take tremendously large forces to slide blocks past each other and create earthquakes. Seismologists could find answers to these questions only by analyzing new instrumental observations of earthquake sources.

Analyzing Earthquake Sources

Only a few of the largest earthquakes near the surface actually rupture the ground, and neither deep earthquakes nor submarine earthquakes can be observed directly. Thus, the analysis of an earthquake's source can generally only be accomplished using indirect observations. Just as we study distant stars through their brightness and spectral pattern, we determine the kinematic and dynamic parameters of the rupture process and the forces generating it by using the seismic waves radiated by the source. These waves are recorded as seismograms at seismic stations.

Seismograms contain information about the source process. However, their use has certain limitations and compromises. The shape of a seismogram is not only deter-

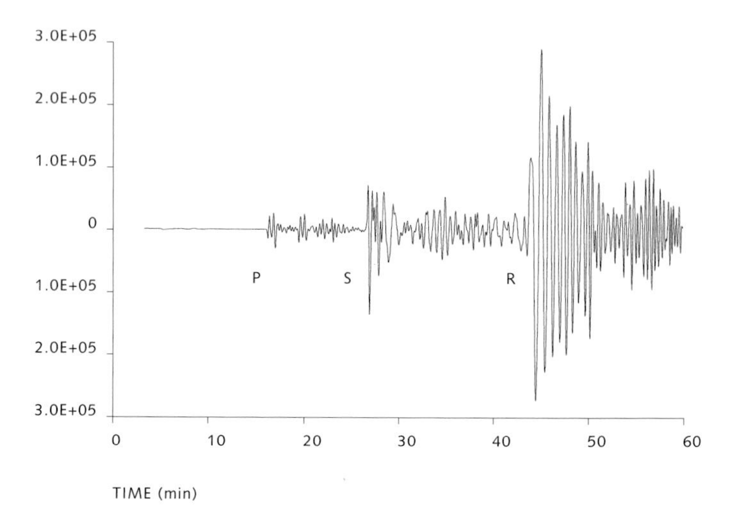

Seismogram, recorded in Stuttgart, Germany, of a magnitude-7.1 earthquake from 100 kilometers depth under the Japanese island of Hokkaido on January 15, 1993. The horizontal axis gives the time in minutes. The beginning of the recording is the source time. The amplitude is the vertical component of the ground velocity (given in nanometers per second) between periods of 20 and 120 seconds.

The first wavegroup arriving (P) is a compressional wave. S designates the shear wave. R is the beginning of a train of dispersed surface waves (Rayleigh waves).

mined by the source process, but is also influenced by the propagation path of the waves between the source and the receiver. Astronomical observations made from Earth have similar problems. The passage of light through the atmosphere distorts the image of an object. Fortunately, we now have a good picture of the structure of the Earth. And, in contrast to the atmosphere, it does not vary with time. In many cases, we can correct the seismograms for the influence of the propagation path.

However, the study of earthquake sources has a serious problem. The physical parameters that determine the course of the source process—the energy, dislocation, mechanical stress, rupture velocity, and rock parameters—are related by complex, nonlinear laws. Nonlinear behavior means that under certain circumstances, nearly infinitesimally small changes in the initial or boundary conditions of some parameter can completely change the course of the process. Thus, factors that appear to be unimportant can dramatically change the course of the source process in time and space. That is why no two earthquakes are exactly alike. Each earthquake is a unique occurrence that will not be repeated.

The investigation and description of earthquake sources is based mainly on model calculations. First, the earthquake source is simulated mathematically with a simple mechanical model. With the help of theories about the propagation of elastic waves and knowledge of the structure of the Earth, we calculate the ground motion that would result from the forces and deformation that the model earthquake source exerts on its surroundings. The result is a synthetic seismogram. The parameters describing the model are then changed until the synthetic and the recorded seismograms are most similar. This method is called the *parameterization of the earthquake source.* The parameters are used to classify the source and to distinguish between sources that are from different tectonic regions and depths. They should agree with values determined from earthquake measurements or from laboratory rupture experiments.

Earthquakes in the Laboratory

For a long time, seismologists received only scant support for their work in source mechanics from the engineering sciences. Engineers are more interested preventing a fracture than in the rupture process itself. Thus, detailed investigations of rupture dynamics only began in the past two decades. Fracture experiments present an opportunity for learning how to interpret the source mechanics from seismograms by using analogies.

Mechanically, fractures can be coarsely separated into two types: *extension cracks* and *shear cracks.* When an extension crack develops, the resulting surface is a free surface and is therefore free of forces. If a glass plate falls, the breakage can be considered extension cracks. The pressure deep inside the Earth from the rocks above prevents the development of free surfaces. Extension cracks can only develop as cracks or fissures close to the surface of the Earth. They are usually not sources of earthquakes.

We can investigate shear cracks in the laboratory. A paving stone of gneiss is clamped in a vise to prepare for an experiment. In crystalline gneiss, freshly cut surfaces often show parallel bands. It is likely that the rock will slide more easily along these bands than perpendicular to them. We place the rock in the vise so that the layers form an angle of 45 degrees with the axis of the vise and place the vise in an oil bath. Using a hydraulic system, we can adjust the hydrostatic pressure in the oil bath to simulate the ambient pressure in the Earth. Because rocks deform plastically (like a very viscous fluid) when changes in the forces are very slow, it is reasonable to assume that the hydrostatic pressure is like the lithostatic pressure—the pressure in the Earth caused by the load of rocks above. Now, as an analogy to the increase in tectonic stresses, we slowly close the vise. The stresses on the rock caused by the vise are superimposed on the omnidirectional hydrostatic pressure. The block is compressed along its axis. Following simple laws of mechanics, the maximum shear stress develops at an angle of 45 degrees to the stresses caused by the vise. According to Newton's Third Law, "for every action, there is an equal and opposite reaction." This means that rock deforms until the internal counterforces are exactly equal, but are opposite to these two stresses. The rock is then in internal equilibrium.

We know from experience that if the force exerted by the vise on the sample exceeds a certain level, the sample will experience a sudden deformation that we call a rupture. Since the shear stress reaches a maximum along A–A′, the shearing motion will begin there. In 1773, the French physi-

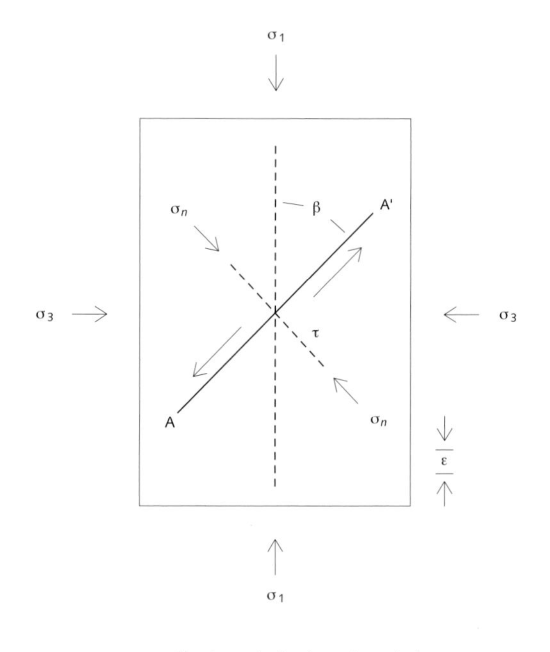

$$\sigma_n = \tfrac{1}{2} \times (\sigma_1 + \sigma_3) - \tfrac{1}{2} \times (\sigma_1 - \sigma_3) \times \cos(2\beta)$$
$$\tau = \tfrac{1}{2} \times (\sigma_1 - \sigma_3) \times \sin(2\beta)$$

Simulation of an earthquake source. Components of mechanical pressure and shear stresses in a test sample:

σ_1: Applied load.

σ_3: Simulation of local, i. e., hydrostatic, pressure.

If the shear stress τ surpasses the shear strength of the rock at the surface A–A′, the rock will rupture and produce a sudden displacement.

cist Charles Augustin de Coulomb defined the criterion for the stability of the upper limit of the shear stress τ:

$$\tau = \tau_0 + \mu \times \sigma_n$$

In this equation, τ_0 is the resistance of the rock to shear forces without the surrounding pressure ($\sigma_3 = 0$), μ is the coefficient of internal friction of the rock, and σ_n is the stress that acts perpendicular to the rupture surface. The equation gives only the critical value for the maximum shear stress. It says nothing about if and when the rupture will continue. The rupture depends on the relationship between stress and strain in the material. When τ exceeds some maximum value, the rock deforms rapidly, which decreases the stress τ. If the rock cannot reach an equilibrium with the acting forces, the system becomes unstable and the rupture process begins. The rupture would propagate forever if the material did not get stiffer with increasing deformation, the so-called *strain hardening*, which leads to a renewed increase in the stress.

The relationship between stress and deformation depends on the behavior of the material and can vary considerably. Inhomogeneities in the rock, unhealed surfaces from previous ruptures, water in the pores of the rock, temperature, and pressure can all determine whether the rupture will be a rapidly occurring, seismically effective brittle fracture that causes an earthquake or a creep event that is hardly noticeable.

Whether the rupture remains concentrated in a small volume or expands like an avalanche depends on the unevenness in the distribution of the material's strength and the amount of deformation energy stored in the rock. With the propagation of the fracture, some kinetic energy is lost breaking up the rocks along the developing fault zone and some is radiated away as seismic waves. Below a certain level, there is not enough energy to push the tip of the fracture any farther. A new equilibrium develops between the mechanical forces and the material's strength. The rupture ends.

The Earthquake Focus as a Point Source

The simplistic representation of the earthquake focus as a point source of kinetic energy is not realistic because the source would have an infinite energy density. It is, however, a good enough model to allow the quantification of several fundamental parameters of the source: the location, the origin time, the kinetic energy it radiates, and the equilibrium forces between the source and its surroundings. These parameters are discussed in the following sections.

The Epicenter, Origin Time, and Source Depth

Determining the exact location of an earthquake is not as easy as you might think. A source process must begin at some microscopically small place, practically on a molecular

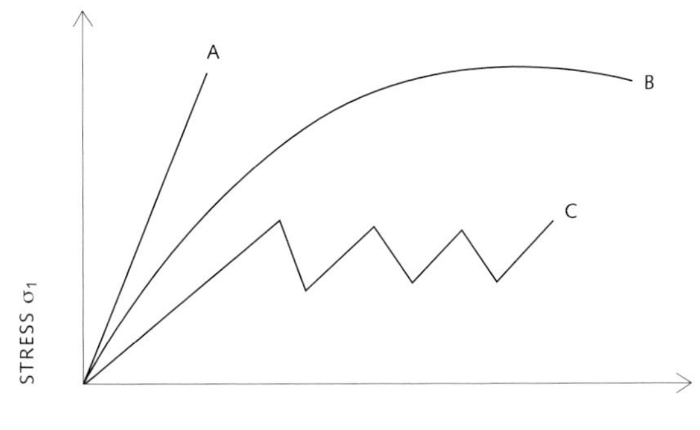

Relationship between mechanical load (stress) and deformation for different types of rock before rupture begins:

A. Mechanically brittle rocks, typical for granites in the upper part of the Earth's crust.

B. Under greater load, the rock is ductile; before it can rupture, the material flows. Typical for rocks at high temperature; for example, gneiss in the lower crust.

C. During the source process, sticking and slipping alternate. During slipping, the dynamic friction is low, and the material stiffens and static friction increases, slowing the rupture until it sticks. The stress increases again until the rupture resumes. This is called *stick-slip motion*. It is typical of brittle rock in a heterogeneous rock matrix. The rate of shear dislocation changes rapidly, producing a series of strong acceleration pulses with dangerous consequences for buildings.

scale. If it were to begin at two or more points at the same time, it would be associated with the transfer of information at an infinite speed. This is not possible because the speed of sound in the medium limits the propagation, at least at the low level of energy at the beginning of the rupture. The geographical coordinates at which the fracture begins are called the *epicenter.* If we also know the source depth, we call this location the *focus* or *hypocenter.* These terms were introduced by Robert Mallet, an Irish engineer, when he investigated an earthquake that occurred near Naples, Italy, in 1857.

A minimum amount of energy must be radiated by an earthquake for it to be noticed and for its focus to be located. This energy cannot come from a point source. Only after the rupture has achieved a certain size does it radiate enough energy as seismic waves, so that a first group of waves can be recorded by a seismograph station. This first wavegroup is commonly called the *primary wave,* or *P-wave.* It is made of compressional waves, the type of wave that propagates fastest and therefore has the shortest travel time from the origin to the receiver.

We can also use the arrival times of later wavegroups, especially the secondary waves, or S-waves, to determine the source of an earthquake. S-waves are shear waves. Using the arrival times of P- and S-waves at several seismograph stations, we can easily locate the source of the quake. To do this, we need to know the velocities at which the waves propagate inside the Earth. The velocities, however, are not constant but depend mainly on the depth to which the waves penetrate the Earth. Over the past several decades, improvements in both the determination of earthquake origins and the travel times between the origins and the seismograph stations have provided a precise model of the distribution of propagation velocities inside the Earth. With this knowledge, the determination of the source location is reduced to a navigation problem. The following quantities must be determined: latitude and longitude, depth, and origin time. At least four seismograph stations must record the arrival times of the waves in order for us to be able to solve the system of equations. In the case of strong earthquakes, which may be recorded at hundreds of stations, the amount of data reduces the error in the calculation. Today, we can often determine the focus of an earthquake and its origin time with precisions of better than 1 kilometer and one-hundredth of a second, respectively.

Intensity and Magnitude

The effects of an earthquake's shaking on man-made structures and the landscape are described by *macroseismic scales.* The 12-level scale for shaking intensity introduced by the Italian seismologist Giuseppe Mercalli in 1902 has been modified and updated for modern building practices. Today, shaking intensity is usually measured with the *Modified Mercalli Intensity Scale* (MMI Scale). It was developed in 1931 by the Americans Harry Wood and Frank Neumann. In abbreviated form, its levels are:[3]

1. Not felt, except by a very few under especially favorable circumstances.
2. Felt only by a few persons at rest, especially on upper floors of buildings.
3. Felt quite noticeably indoors, especially on upper floors of buildings, but many people do not recognize it as an earthquake.
4. During the day felt indoors by many, outdoors by few. At night, some awakened.
5. Felt by nearly everyone; many awakened. Some dishes, windows, etc., broken.
6. Felt by all; many frightened and run outdoors. Some heavy furniture moved. Damage slight.
7. Everybody runs outdoors. Damage negligible in buildings of good design and construction; slight to moderate in well-built ordinary structures; considerable in poorly built structures.
8. Damage slight in specially designed structures; considerable in ordinary substantial buildings with partial collapse; great in poorly built structures. Fall of chimneys, columns, monuments, walls.
9. Damage considerable in specially designed structures; well-designed frame structures thrown out of plumb; great in substantial buildings, with partial collapse. Buildings shifted off foundations. Ground cracked conspicuously.
10. Some well-built wooden structures destroyed; most masonry and frame structures destroyed with foundations; ground badly cracked. Rails bent. Shifted sand and mud.

11. Few if any (masonry) structures remain standing. Bridges destroyed. Broad fissures in ground. Earth slumps and land slips in soft ground.
12. Damage total. Waves seen on ground surfaces. Objects thrown upward into air.

After strong earthquakes, the effects are determined in field studies using questionnaires. Each city or neighborhood is assigned an average level of intensity. The intensity levels are usually given a geographic representation as isoseismal lines on intensity maps. In earlier times, the size of the earthquake was given by the maximum intensity. Despite the often rather subjective estimation of the degree of shaking, these maps represent an important basis for planning earthquake-resistant buildings, particularly when historical earthquakes are included in the evaluation.

The MMI Scale, based on observed damage, is not quantitative enough for modern seismology. In addition, many earthquakes occur in regions with little or no population. In 1935, the American seismologist Charles Richter suggested a procedure that can be used to calculate the seismic energy released in an earthquake by using the amplitude of the ground motion measured with seismographs. The ground motion of a large earthquake can be thousands or even millions of times larger than that of a small earthquake. Richter therefore introduced a magnitude classification similar to the astronomical classification of the brightness of the stars. His earthquake magnitude is calculated from the logarithm of the maximum ground amplitude at a predetermined distance from the focus. This gives an objective quantification of the strength of an earthquake. The value is called the *Richter magnitude*, or *local magnitude*. It can be related empirically to other parameters describing an earthquake, such as the intensity or the source depth. One often-used equation relates the Richter magnitude (M) to the kinetic energy (E) released by the source:

$$\log E = 1.5 \times M + 4.8$$

where E is given in joules and log is the base-10 logarithm. Richter's formula for magnitude is valid for earthquakes of lower magnitude in the upper crust recorded at local seismograph stations. As the magnitude increases, the extent of the source also increases. However, Richter's formula is not strictly valid when $M = 6$ or more. The magnitudes of stronger earthquakes, deep-focus earthquakes, or more distant sources can be determined using alternative procedures. One frequently used magnitude value based on the amplitudes of surface waves taken from certain frequency, or period, bands is called the *surface-wave magnitude* scale. One preferred period band is around 20 seconds. Another magnitude scale is based on the amplitudes of body waves (*body-wave magnitude*). It is mostly applied for determining the size of deep earthquakes, where the excitation of surface waves is small. The *moment magnitude*, a fourth-magnitude value, is based on the concept of the seismic moment (see page 49). It is largely used for the description of very strong earthquakes.

Forces and Displacements

In 1923, the Japanese seismologist Hiro Nakano published a fundamental contribution to *seismotectonics*—the description of the relationship between the forces and displacements in the earthquake focus and its tectonic setting. Early in the analysis of seismograms, observers noticed that the first motion from the source could be either away from or toward the source. For some seismologists, this was the proof that Reid's model is not generally valid. It seemed plausible that if the first motion pointed away from the source (a compression), it would come from an explosion, while first motion toward the source (a dilatation) would imply a cavern collapse. Nakano calculated the expected radiation diagrams for different combinations of forces. By comparing them with seismograms, he found that if he replaced the earthquake focus with a pair of forces working in opposite directions along an imaginary surface, he could produce a simple distribution of compression and dilatation. Such changes in polarity had been observed, but since no one had a reasonable interpretation for them, they were assumed to be measurement errors.

In mechanics, the pair of force corresponds to the shear forces necessary to generate an earthquake. With Nakano's formulas it was finally possible to project the results from the analysis of seismograms backward to determine the position of the rupture surfaces in the Earth and the movement along them. With this simple method it is, however, not possible to immediately distinguish between the actual slip or fault plane and another plane perpendicular to it. A more complex calculation of the direction in which the source extends is required to separate them. The orientation of the tectonic stress field that is associated with the earthquake, on the other hand, can be uniquely determined using *fault-plane solution* methods. These techniques are among the most important pillars of geodynamics.

The Spatial Extent of the Earthquake Source

Fault Length and Rupture Area

On maps of their distribution, earthquakes are usually displayed as points. This simplification is often useful, but nonetheless fictitious. The maximum density of energy that can be stored in the rock—the energy of deformation per unit volume—is limited by the strength of the material. Since an earthquake draws its energy from the source volume, a source with a greater magnitude must take up a larger volume.

The source volume does not have a unique definition. It depends, among other things, on the methods used for measuring it. If it is determined from the analysis of seismograms, the source volume is given by the sum of the regions that radiate kinetic energy during the rupture and thereby generate shaking at the Earth's surface. The definition of the

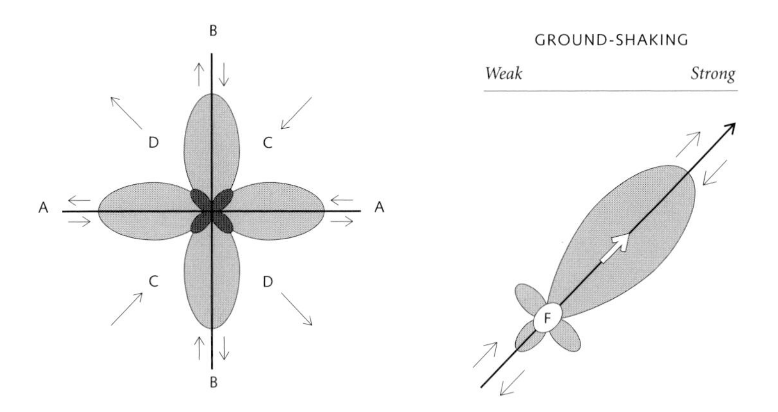

Left: The radiation pattern of seismic waves from a point source at the center of the figure. The slip may correspond either to the dislocation shown along line A–A or that along B–B. Both assumptions give the same radiation pattern. The diagram shows the amplitude distribution of the compressional or P-waves (dark-gray) and the shear or S-Waves (gray). The projection gives an example of what the radiation would be for a strike-slip fault like the San Andreas Fault (along B–B), seen from above. The maximum amplitude of the shear waves is four times that of the compressional waves. The first motion of the compressional waves divides the radiation pattern into four quadrants, two with compression (C) and two with dilatation (D). From this distribution, we can draw important conclusions about the tectonics in the region of the source.

Right: The radiation pattern of the shear waves for an extended source. The plane of the projection is perpendicular to the surface of the Earth. The slip direction, indicated by black arrows, corresponds to a thrust fault. From the focus (F), or hypocenter, the rupture propagates toward the surface (white arrow). In this direction, a wave of larger amplitude develops due to interference. It can cause very strong ground motion. The unusually strong ground motion observed during the Northridge, California, earthquake (January 17, 1994) was probably caused by such effects.

source volume from tectonics depends less on the shaking and more on the ground deformation that remains when the shaking has ended. Sometimes earthquakes happen almost aseismically: Over the course of several hours, creeping occurs along the fault and produces dislocations, but it cannot be measured with seismometers. The changes can be measured at the Earth's surface geodetically or by using tiltmeters. Recently, such changes have also been observed by analyzing radar satellite data.

The source volume is usually limited to a narrow, bandlike zone. For example, dislocations from the 1906 San Francisco earthquake were observed only in a 5-meter-wide strip along several hundred kilometers of fault. There was no lasting deformation a few hundred meters to the left and right. This concentration of movement in a narrow zone is the result of a self-reinforcing process. The rock that has been shattered and faulted by an earthquake is weaker than its surroundings. Future earthquakes will therefore preferentially rupture this zone. Over the course of time, the fracture zone becomes narrower, in some cases almost to the point of becoming a slippery track in the Earth. Such tracks can extend over hundreds of kilometers. In the mathematical description of the source process, we often substitute the more easily described, two-dimensional fault plane for the long, narrow, three-dimensional source volume. Occasionally, we use only a one-dimensional description of the source, the fault line.

The following empirically determined equations describe the approximate relationships between the magnitude (M) of a source and its area (A in cm^2) and its linear extent (L in cm):

$$\log A = 1.02 \times M + 6.0$$
$$\log L = 0.5 \times M + 3.2$$

where log is the base-10 logarithm. These equations are valid for earthquakes in the Earth's crust with magnitudes greater than 4.

In quarries or cliffs, we can often recognize places where blocks of rock moved relative to each other in past geological eras. In such exposures, we find layers of strongly stressed rocks ranging from centimeters to meters in thickness. They are sandwiched between sections of undisturbed rocks. Sometimes, the altered rocks have only been polished by the movement. However, if the deformation has taken place over a long period, the rock in the interface has been ground into powder, that is, it has been *mylonized.*

The growth of an earthquake from a tiny source to more than 1000 kilometers length is an extremely complex, dynamic process. Even the largest earthquake begins in a microscopically small region. When does such a micro-event grow into a macro-quake? If we have the right microphone, we can hear continuous crackling in any cave dug into rock. The mechanism producing these mainly ultrasonic acoustic emissions corresponds to an earthquake with a source only fractions of a millimeter long. One necessary condition for the avalanche-like growth of such microcracks is a mismatch between the forces acting on the rock from outside and its ability to diffuse these forces through internal elastic deformation. If the reaction through deformation described by Newton's Third Law can be equal to the action of the outside forces, the system will be stable. If the outside forces exceed the rock's ability to deform, the rock will break.

Heterogeneities in the geological and mechanical structure of the Earth mean that the equilibrium between these two forces varies with location. Sub-millimeter-sized pores and cavities can reduce the strength of rock by orders of magnitude. In such cases, the conditions for instability are easily satisfied, but the break is usually limited to a tiny section of rock. The result is an acoustic emission, not an earthquake. A rupture can only propagate over a larger region when the zones of stability and instability are properly distributed in the potential source region. Enough contiguous or neighboring regions of instability must be present so that the rupture has the momentum to overcome zones of stability.

The propagation of a rupture is also determined by the change in the strength of the material with movement along the propagating rupture front. We know from experiments that the resistance of the rock to shear forces is mainly based on static friction between neighboring rock particles. Macroscopically, the pressure on the crustal blocks in the source region determines the amount of static friction countering the outside forces. When the rupture starts, the high static friction is replaced by low dynamic friction. This is equivalent to a reduction in the strength of the material—the rupture can propagate more quickly. A larger rupture means a larger dislocation along the fault plane. This also means greater compression of the material and therefore stiffening, with the resulting increase in the strength of the material. In this interplay, *stick-slip faulting* develops, in which the complete source process is made up of a series of individual ruptures. The squeak that we sometimes hear when writing with chalk on a chalkboard is the result of a similar sequence of interruptions.

The dependence of rupture propagation on many individual factors leads to a surprisingly simple relationship between the number of earthquakes which occur and their magnitude. This relationship was first formulated by the American seismologists Beno Gutenberg and Charles Richter:

$$\log N = a - b \times M$$

where N is the number of earthquakes with magnitudes greater than M, and a and b are constants for a given region during a given interval. The log is the base-10 logarithm. The number of earthquakes decreases logarithmically with increasing magnitude. For the entire world, a is in the range of 7 to 8, and b is 0.9. On the average, there is one magnitude-8 earthquake every year, about 100 with magnitudes greater than 6, and more than 50,000 with magnitudes greater than 3.

The b-value is related to the heterogeneity of the rocks in which the earthquake occurs. If they are more fractured, the b-value is larger. Thus, the b-value is used to characterize a region seismotectonically. Scientists also try to relate changes of b over time to changes in seismic activity before or after a large earthquake. The a-value is a constant that stands for the seismic activity rate.

Slip, Slip Velocity, and Rupture Velocity

The main characteristic of an earthquake source is the rapid displacement—the slip or the shear dislocation—that develops along the rupture plane. Such dislocations only break through to the Earth's surface when the earthquake is very large, usually of magnitude greater than 7. The largest known dislocations, more than 10 meters, occurred in 1897 during the Great Assam Earthquake in northern India.

At each point in the source volume, a dislocation develops during the course of an earthquake. The amount of slip can vary from zero to some maximum value, D. Experience has shown that we can relate D to the total length of the rupture, L. The empirical rule is that $D = 10^{-5} \times L$. A rupture 10 kilometers long would have a maximum dislocation of about 10 centimeters.

Like rocks sliding from a pile, the movement at the starting point of the earthquake, its focus, triggers dislocations around it. The velocity at which the rupture front propagates is called the *rupture velocity*. It and the time derivative of the dislocation or slip, the *slip velocity*, are important parameters for the mathematical description of the earthquake source. These two quantities contribute to the shape of the recorded seismograms.

In addition to calculating the focus and magnitude from the seismograms, the seismologist must use the recordings and model calculations to determine the size and direction of kinematic source parameters, such as rupture area, slip, slip velocity, and rupture velocity. The slip velocity is usually less than 1 meter per second, while the rupture velocity is several kilometers per second. By interpreting these quantities using knowledge of the rupture mechanics of brittle and ductile rock, it is possible to draw conclusions about dynamic processes within the Earth that we cannot observe directly. The seismic waves often have larger amplitudes in the direction that the rupture propagates. This directionality is caused by the superposition of waves that have been radiated from different sections of the slipping fault and are at least partially in phase with one another. The amplification is strongest when the rupture velocity is the same as the propagation velocity of the seismic waves.

The seismic moment, M_0, is also important as a characterization of an earthquake source. It has the advantage that it can be determined without the unwieldy and time-consuming process of forward-modeling and comparing the results with seismograms. Rather, it can be calculated directly from seismograms using mathematical inversion techniques:

$$M_0 = \mu \times A \times D$$

where μ is the shear modulus (the ratio of the shear stress to the shear strain) of the medium, A is the rupture area, and D is the average slip on the rupture surface. The seismic moment only depends on the final displacement and size of the rupture, not the speed at which it occurs.

Earthquakes That Should Not Exist

Deep Earthquakes

During the evening of June 8, 1994, the occupants of the upper floors of high-rise buildings in Toronto experienced unpleasant, long-lasting swaying. It soon became clear that this could only have been caused by an earthquake. However, eastern Canada only experiences occasional, weak earthquakes that are rarely felt, since it is far from a plate boundary. Even experienced seismologists were surprised to find that the waves causing the swaying were from an earthquake that occurred more than 6000 kilometers away. The magnitude-8.2 earthquake with a focus 620 kilometers below Bolivia was the strongest earthquake ever recorded from such a depth.

When Herbert Hall Turner, a seismologist at Oxford, calculated earthquake source depths of several hundred kilometers in 1922, he was very skeptical. He assumed that he had made some systematic error in his calculations. His mistrust of his results was shared by a leading earth scientist of the time, Harold Jeffreys of Cambridge, England.

There are important theoretical and experimental considerations that seem to exclude the likelihood of earthquake sources in the hot, ductile rocks of the Earth's mantle. The pressure at great depths implies very high coefficients of friction between grains of rock. Thus, unimaginably high shear forces would be needed to cause a dislocation equivalent to that of a near-surface earthquake of similar magnitude. In addition, the plasticity of the mantle rock should prevent the buildup of shear forces.

However, after investigations in 1930 by Japanese seismologist Kiyoo Wadati, no one could deny that there were earthquake sources below the cold and brittle crust. Several scientists sought an alternative explanation and suggested source mechanisms other than shear fractures. Current hypotheses for the cause of these deep earthquakes include rapid volume changes associated with phase changes in the rock. These could be related to instabilities in the crystal structure of the deep rocks as the pressure and temperature change. However, all seismograms, even from the deepest earthquakes, can be explained by a shear dislocation.

In 1934, Wadati recognized a pattern in the foci of earthquakes occurring between 50 kilometers and several hundred kilometers depth, marking what we call a *subduction zone* in today's plate tectonic terms. A subduction zone is where the lithospheric plate is bending and descending into the Earth. The cold, water-saturated rocks in the upper part of the plate are drawn downward by gravity, and the lower side is pressed against the more solid rocks of the lower mantle. The poor thermal conductivity of the subducting plate prevents its interior from heating rapidly, so that shear stress cannot be released immediately by plastic deformation. Compression due to the lithostatic pressure only depends on the depth. Thus, it cannot be employed to contradict the claim that, even in a subduction zone, rock friction or strength resists the shear forces. Various theories attempt to reconcile these problems. The dehydration of the water-rich lithospheric rocks during heating could play a role. As the rocks become dehydrated, the water pressure in their pores increases, while the effective surrounding pressure decreases. This reduces the friction along a potential shear surface. On the other hand, the water can also act as a lubricant.

Many questions about the mechanisms of deep earthquakes remain open today. In Europe, deep earthquakes occur in southern Spain, in the Tyrrhenian Sea, and in the Carpathian Mountains. It is difficult to relate any of these earthquakes to a subduction zone.

Intraplate Earthquakes

About 99 percent of the energy released by earthquakes is released at plate boundaries, with more than 90 percent being released at convergent boundaries, where two plates collide. This observation is fundamental support for the plate tectonics theory, which predicts that most deformation will occur at plate boundaries and not in their interiors.

As with every model, however, plate tectonics is only an approximate description of reality. For example, the earthquakes that occur in the northeastern U.S. are far from any plate boundary. The 1755 earthquake near Boston, the Cape Ann earthquake, is estimated to having had a magnitude of 6.2. Such earthquakes are considered to be the effects of local stress concentration or zones of weakness due to geological heterogeneities in the Earth's crust. Although they are rare, extremely strong earthquakes with devastating conse-

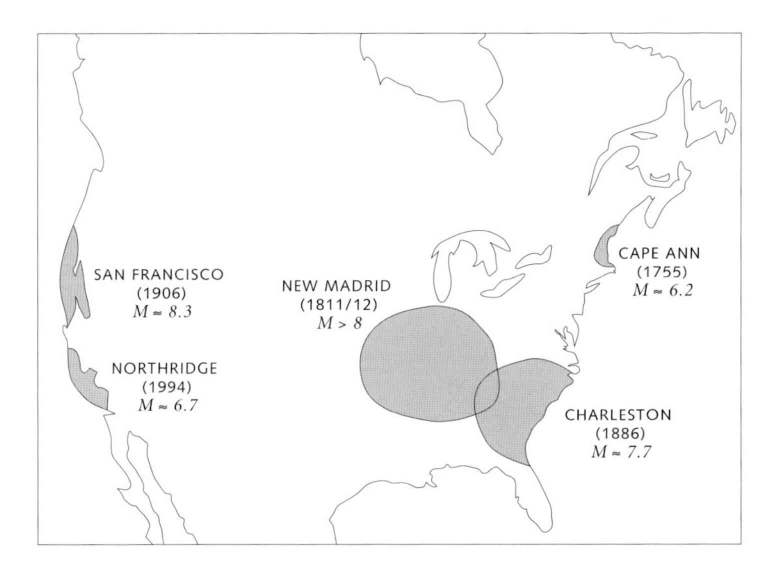

Comparison of large earthquakes in the eastern, central, and western U.S. In addition to the name, magnitude (M), and year of occurrence, the map shows the regions affected by the earthquakes. These are the regions where the intensities of ground-shaking reached or exceeded 6 on the MMI-scale. Note the large regions affected by the New Madrid and Charleston earthquakes.

(Map adapted from information from the U.S. Geological Survey.)

quences have been known to occur in the interiors of the plates. For example, China's earthquake catalog, which covers 3000 years, includes strong earthquakes with epicenters more than 1000 kilometers from either the collision zone of the Himalayas or the subduction zone of the Pacific plate. On the African plate, Libya's seismic activity was minimal for years. However, in 1935, Libya was shaken by four earthquakes with magnitudes of 7 or greater within a few weeks.

The largest intraplate earthquake in modern times occurred in the 1800s in the American Midwest. In late 1811 and early 1812, three earthquakes with magnitudes greater than 8 shook the Mississippi River valley where it borders Kentucky and Missouri near New Madrid, Missouri. Their magnitudes are similar to the strongest earthquakes experienced in California. It is still not clear how the necessary stress could build up in a region that is considered to be tectonically stable. Some speculate that the earthquakes are related to the huge buildup of sediments from the Mississippi River and the resulting loading of the Earth's crust. It is also possible that they were caused by the transfer of stress from the Atlantic and Pacific plate boundaries. In the plate interior, the compression rate (less than 1 millimeter per year) is much less than at the plate boundaries. Over thousands of years, however, it would build up to produce an unstable state of deformation in a large region.

When the length of a seismic cycle exceeds historical knowledge of the seismicity in a given region, the inhabitants are surprised when an earthquake occurs. Intraplate earthquakes usually recur only over long periods of time and are therefore particularly dangerous because people are not expecting them.

The Seismotectonics of California

California is a part of the Circumpacific Belt, and as part of an active plate boundary, displays all the features of plate tectonics. The best-known tectonic structure is the San Andreas Fault. Like a long trough it extends through most of western California. Andrew Lawson mapped the fault at the end of the nineteenth century. He named it after a reservoir south of San Francisco. The fault achieved worldwide recognition on April 18, 1906, when it ruptured in the great San Francisco earthquake. From San Juan Bautista, 50 kilometers south of San Jose, to Point Arena, almost 300 kilometers to the northwest, displacements of up to 6 meters were recorded wherever the fault crossed streets, rivers, fences, or rows of trees. In all cases, the western side of the fault moved north and the eastern side moved south.

In the language of plate tectonics, the San Andreas Fault is a *transform fault*—a fault in which two plates slide past each other along a vertically oriented fault plane. The rupture surface extends nearly 20 kilometers into the Earth. Over a distance of 1300 kilometers, it connects two separate sections of divergent boundaries: the opening of the East Pacific Rise in the Gulf of California and the Juan de Fuca Ridge off the Pacific coasts of Oregon, Washington, and British Columbia.

The San Andreas Fault separates the North American plate from the Pacific plate. The plates are moving past each other at a rate of about 5 centimeters per year. The direction of horizontal motion corresponds to the dislocation observed in the San Francisco earthquake. Seen from space, the San Andreas Fault is a dominant feature. It seems to be the only plate boundary. At the Earth's surface, however, the transition from one plate to the other is more complex. In some places, like Hollister, southeast of San Jose, the fault is very narrow. There, it is possible to stand with one foot on the North American plate and the other on the Pacific plate. At other places, such as Tejon Pass or El Cajon, plate motion is accomodated by many parallel faults. In the San Francisco Bay area, only 40 percent of the plate motion occur at the San Andreas Fault. Most of the rest is accomodated by the Hayward and Calaveras faults. Northeast of Los Angeles, the San Andreas Fault bends toward the west, which is clearly related to the strong north-south compression in this region.

How do narrow fault zones develop? Between the surfaces of the fault are zones with low shear strength. They contain mylonite rock—rock that has been crushed—and can be several centimeters to 1 meter thick. The zones separate the opposing faces of the crustal blocks and drastically reduce the friction between them. In part, the zones date back to the time when the faulting originally began. As a result of mechanical loading and the rheological behavior of the rock, creeping occurs along these shear zones at rates of centimeters to meters per century. This plastic creeping counteracts the healing and strengthening of the opposing faces of the disturbance and encourages the development of new fractures. We can imagine this as a feedback process, where what begins as a three-dimensional web of small cracks and fractures becomes a two-dimensional surface through self-organization and self-amplification.

The San Andreas Fault is made of many individual segments. Each of the segments behaves differently relative to the overall plate motion. Creeping in some sections alternates with segments where the faces of the plates seem to be firmly stuck. The differences are apparent in the maximum magnitude of the seismic activity: The more firmly the plates are stuck, the more potential energy can collect before the next rupture. The varying shear strength along the fault is not surprising. Since the creation of the fault, its opposing faces have moved several hundred kilometers relative to each other. This means that completely different types of rocks can lie opposite each other along the fault.

The Pacific plate will continue to move past the North American plate for the next several million years. People will only notice this movement for a few seconds or minutes when an earthquake happens. The transition zone between the lithospheric plates can be soft or hard. If the transition zone is soft, then movement will lead at most to a quiet crackling, with many small earthquakes. The segment just south of the San Francisco Bay area, between San Juan Bautista and Parkfield, is such a region. On the other hand, the transition zones are hard near San Francisco and Los Angeles, as the magnitude-8 earthquakes of 1906 (San Francisco) and 1857 (Fort Tejon) demonstrate. When there

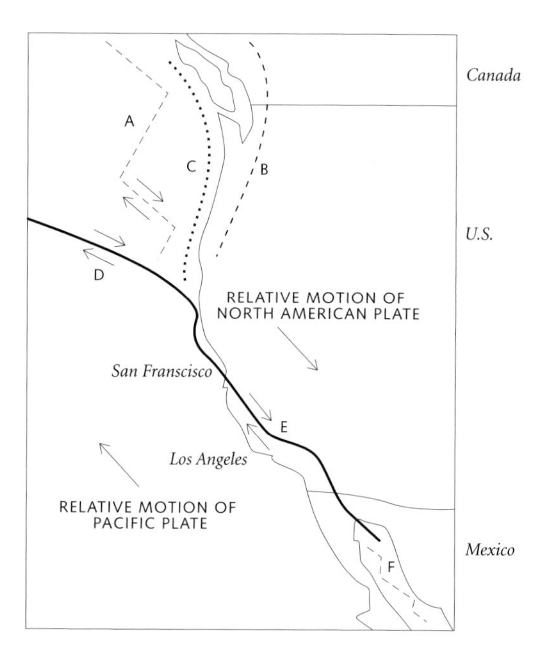

A *Juan de Fuca Ridge*
B *Cascade Range*
C *Subduction Zone*
D *Mendocino Fracture Zone*
E *San Andreas Fault*
F *East Pacific Rise*

The San Andreas Fault in California represents an important, nearly vertical interface between the opposing motions of the North American plate and the Pacific plate. The fault connects the spreading centers of the East Pacific Rise and the Juan de Fuca Ridge. In plate tectonic terms, the San Andreas Fault is a transform fault. Just east of the Juan de Fuca Ridge is a subduction zone that leads to volcanic activity in the Cascade Range.

(Map adapted from information from the U.S. Geological Survey.)

is no seismic activity along a segment of the fault for a long interval, we speak of a *seismic gap*. It is likely that the plate motion at such a section has gotten stuck and the region is becoming stressed, as described in Reid's theory. It is certainly a warning that a large earthquake will occur, but not much more than that. The conditions in the Earth are so complex that we cannot derive a practical method for short-term earthquake prediction from such observations.

Three California Earthquakes

Between 1989 and 1994, three major earthquakes occurred in California. Although they were nowhere near the dreaded Big One in strength and size, they caused 112 deaths and more than $20 billion in damage. Important earthquakes are named after the geographic location of their epicenter and the year of their occurrence. These three earthquakes are Loma Prieta 1989, Landers 1992, and Northridge 1994. They show the diversity of the tectonic activity in California.

Loma Prieta, October 17, 1989

The epicenter of this earthquake lay in the Santa Cruz Mountains, about 100 kilometers southeast of San Francisco. The rupture began at a depth of 18 kilometers. It propagated to the northwest and toward the surface along the San Andreas Fault, until it reached a depth of 6 kilometers and a maximum length of 40 kilometers. Thus, there were no shear dislocations observed at the surface. The magnitude, determined from surface waves, was 7.1—the strongest earthquake along the San Andreas Fault since 1906. It is interesting to note that this earthquake reruptured the southernmost 40 kilometers of the 1906 earthquake. The horizontal displacement was about 1 meter in the usual faulting direction of the San Andreas Fault. Surprisingly, there was also uplift of about 40 centimeters.

Seismologists expected the Loma Prieta earthquake to occur. They had observed a lack of seismic activity in the upper crust under the Santa Cruz Mountains during the past few decades, while the number of earthquakes in the lower crust remained normal. It was a typical *seismic gap.* There were no foreshocks or other anomalies that could have led to a short-term prediction.

The maximum shaking intensity was VIII to IX on the MMI scale, and the peak ground acceleration was 0.5 to 0.6 that of Earth's gravitational acceleration (i. e., 500 to 600 centimeters per second squared). As is usually the case, the consistency of the ground was an important factor in the effects of shaking. Loose sediments and the notorious bay mud of old arms of the San Francisco Bay exacerbated the effects of shaking by one or two units on the MMI scale compared to nearby bedrock sites. In the Marina District of San Francisco, the damage due to liquefaction of the ground was particularly strong. The similarity in the damage pattern to the earthquake of 1906 is unmistakable.

Landers, June 28, 1992

Landers is a small town in the Mojave Desert, about 170 kilometers east of Los Angeles. From Landers, the rupture propagated 70 kilometers to the north–northwest along several faults that run nearly parallel to the San Andreas Fault. The earthquake had a magnitude of 7.6 and was followed three hours later by the Big Bear earthquake, 35 kilometers to the west with a magnitude of 6.5. Along almost its entire length, the horizontal displacements were in the usual faulting direction of the San Andreas Fault. Across the rupture zone, which was 100 meters wide in places, the measured dislocations were up to 5 meters.

The main shock was followed by thousands of aftershocks. There is evidence that the Landers earthquake induced seismic activity over large areas of the western U.S. The increase in seismicity was particularly strong in geothermal fields and regions with recent volcanic activity up to 1000 kilometers from the epicenter.

Northridge, January 17, 1994

The focus of this earthquake was located directly under Northridge, a city in the San Fernando Valley about 30 kilometers northwest of Los Angeles. The rupture began at a depth of about 20 kilometers and propagated upward until it stopped about 7 kilometers below the surface. The source mechanism of the earthquake was *thrust*, or vertical, on a fault plane tilted westward with an angle of 35 to 40 degrees from the horizontal. Thus, the orientation of the fault plane and the type of dislocation are fundamentally different from earthquakes occurring along the San Andreas Fault. Thrust faults are common along the north flank of the Los Angeles Basin. They are caused by compressive forces acting in the north–south direction. These lead to a shortening of the crust of up to 7 millimeters per year. The rapidly growing mountains of the Transverse Ranges, in which the San Fernando Valley lies, are a visible expression of these forces.

With a magnitude of 6.7, the Northridge earthquake was much smaller than the Loma Prieta and Landers earthquakes. Still, measured by the damage, it was second only to the San Francisco earthquake of 1906. Several factors contributed to this. In thrust faults, stress presses the crustal blocks on either side of the fault together. Friction increases, so that very large shear stresses can develop before the rupture begins. The orientation of the rupture plane is another factor that contributed to the large amount of shaking at the surface. The rupture plane was oriented so that the maximum amount of seismic energy was radiated straight up. The rupture also propagated upward, and the waves traveling upward were amplified by the interaction between the rupture velocity and wave propagation velocity.

Earthquake Prediction

No other question is as controversial among seismologists as whether it is possible to predict the place, time, and strength of an earthquake. Opinions range from, "with enough money, anything is possible," to complete denial. The disagreements are not based solely on scientific arguments. Funding for large, often world-spanning projects in seismology in the past 40 years have been directed toward two goals:

the detection of underground nuclear explosions and earthquake prediction. The first goal has been reached. But the question is still completely open as to whether, and with what methods, earthquakes can be predicted. In addition, the concept of earthquake prediction is so general that it is easy to justify funding by highlighting positive aspects. There is no other area in geophysics in which self-serving optimism is encouraged as much. The media are eager to report successes in earthquake prediction and the supposed relationship between explosions and subsequent earthquakes. Thus, these topics have become a gray area in which it is often difficult even for experts to distinguish between theories from charlatans and valid results from serious scientists. It is also possible that a valid prediction could produce more conflict and damage than an unexpected earthquake.

People commonly imagine that earthquake prediction works like weather prediction. A responsible institution announces that in a certain area on a certain day, there is a certain probability that an earthquake will occur with a magnitude within a certain range. The probability describes the confidence level of the prediction. If it is 50 percent, then the institution believes that there is a one-in-two chance of an earthquake occurring. With our knowledge today, it might be possible to say that during 2002, there will be an earthquake with magnitude greater than 2 in the metropolitan Los Angeles area extending, say, 20 kilometers from north to south and 10 kilometers from east to west. The actual probability for this prediction is greater than 95 percent—an excellent value for an earthquake prediction. But the prediction has no practical value since earthquakes of magnitude 2 cannot be felt and cause no damage.

The example shows how broad the concept of earthquake prediction is. The four parameters—place, time, magnitude, and probability—are related by a sort of uncertainty principle. If we reduce the range for one parameter, the ranges for the other parameters grow correspondingly. If the range for the location is too large—for example, “the Andes” or “California”—then the prediction is useless. If the location is limited to a reasonable area, such as “the region around San Francisco” or “Boston,” and we give a higher magnitude level such as $M > 7$, then the time range must be large. Thus, we must use expressions like “long-term” or “middle-term.” The earthquakes are no longer “predicted” but “foreseen” or “forecast.” While not very specific, such information is nonetheless valuable for constructing buildings and for emergency preparedness measures.

Several years ago, Japanese physicists wanted to use fracture experiments to demonstrate that it is fundamentally impossible to predict the time of the next earthquake with a useful precision. Window-sized glass plates were supported at their edges. The plates were subjected to loads at their centers that increased with time. At some point, the plates broke like the brittle fracture of an earthquake. The interesting result of these experiments was that even though the glass plates were produced using identical methods, the interval between the beginning of loading and the time they broke varied considerably. Microscopically small, normally unimportant regions of inner stress in the glass are responsible for the sudden initiation of the break without any warning. The conclusion is: If the time of breaking cannot be predicted even when the material is homogeneous and the increasing load reproducible, how can we ever predict the beginning of a rupture in the Earth with practical precision when the rock matrix is so heterogeneous and the tectonic stress grows unevenly?

Shortly after the glass plate experiments, the well-known Japanese seismologist Kiyoo Mogi pointed out that if the material is highly heterogeneous, the time at which the material breaks varies much less. If similar experiments had been performed with wood or concrete plates, it would have been possible to predict the time of the fracture more precisely. In addition, the breaking of the wood would have been preceded by audible cracking noises, and ultrasound emissions would have foreshadowed the concrete's collapse. Transferred to the Earth, these conclusions mean that earthquake cycles in heterogeneous regions, like subduction zones, would be more predictable than the cycles of intraplate earthquakes in the more homogeneous lithosphere.

The goal of earthquake prediction research is essentially to find significant precursors—that is, abnormal occurrences that indicate the likelihood of a rupture extending over a large area. For centuries, people have noticed that water levels in wells and groundwater levels are affected by earthquakes. These levels can change not only in response to precipitation, but also with variations in the tectonic stress field. In China and Japan, water levels in deep wells are constantly observed for this reason. The strong earthquake off the coast of Lisbon, Portugal, in 1755 caused oscillations of more than 50 centimeters in the water levels of wells in central Europe. Water levels change because the upper several kilometers of the Earth's crust contain porous and fractured rock that is full of water, just like a sponge. Even the smallest changes in the rock volume produces visible and easily measurable changes in wells.

The seismologist Cinna Lomnitz reports that water level anomalies were observed in China before the destructive earthquake of Tangshan in July 1976. The earthquake killed more than 200,000 people. Several years before it happened, the groundwater level slowly sank in the region. The level dropped more quickly in the months just prior to the earthquake. A few hours before the event, the water level rose so much that several wells overflowed. When the earthquake began, the groundwater rose several meters at many locations.

This report sounds impressive, but it and other reports of earthquake prediction reveal limitations. The observed changes were only recognized *after* the earthquake. There are very few large earthquakes for which, in retrospect, people cannot find some anomalous observation before the event. Such observations are not restricted to instrumental measurements. Seismologists have received numerous telephone calls in which animal owners describe the unusual behavior of their cats, dogs, or birds before an earthquake. Until recently, observations were made on snake farms in China in order to study the thousand-year-old legend that snakes leave their holes in the ground before an earthquake. The observations were stopped after several decades produced no positive results.

In the U.S., Japan, and China, research programs for short-, middle-, and long-term prediction are well funded. The investigations attempt to measure and record, directly or indirectly, factors that are affected by the rupture mechanics of earthquakes. In any rock under lithostatic pressure, there is a continuous creation, closing, and rebuilding of fine cracks that are often only several thousandths of a millimeter long. Such changes cause ultrasound emissions, which can be recorded with highly sensitive ultrasound microphones. We can estimate changes in the stress level of the region by measuring the number and strength of the emissions. Other methods for estimating changes in stress measure the deformation at the Earth's surface geophysically using dilatometers and tiltmeters, or geodetically using triangulation or leveling. In the past few years, modern navigation methods using satellite data allow us to measure large-scale displacements of the ground with a precision of a few centimeters.

There are many excellent methods for acquiring data, but their interpretation is still problematic. Hooke's Law, which relates mechanical tension to deformation, is only conditionally valid in the rocks of the Earth, particularly when the observations span a long time. Creeping can mean that, despite apparently large deformation, only minimal mechanical stresses build up. The relationships in the determination of rock strength are even more complex. The increase in microcracks in rock is connected with increased permeability—that is, water and gases can pass through it. Before an earthquake in Tashkent, Uzbekistan, in 1966, someone noticed an increase in the amount of radon gas dissolved in the water. Radon being mildly radioactive, its concentration can easily be measured with simple methods.

The fundamental problems of trying to predict the time, location, and strength of an earthquake are, however, not instrumental. They probably cannot be solved even with large-scale projects applying expensive methods for measuring and observing. Earthquakes are physically complex processes that never repeat themselves exactly. They can only be quantified with the help of models, which require extensive simplification. Reid's elastic rebound model is a good approximation of the kinematics of an earthquake source. However, it says nothing about the factors that act during the build-up to the earthquake or those that finally contribute to the rupture.

Plate tectonics provides a plausible explanation for the existence of earthquakes. However, these sudden dislocations are not the dominant phenomenon accompanying the motion of continental or oceanic plates relative to each other. Most tectonic displacements and deformations occur in the form of slowly developing plastic flow of material. The visible effects of these movements are only apparent after generations. Earthquakes contributed to the rise of the Rocky Mountains. However, their contribution is only a small fraction of the continuous creeping processes that do not endanger people.

Earthquakes are sometimes viewed as accidents in the tectonic dance of the plates on the Earth. The next earthquake is sure to come, just as the next traffic accident on a freeway. Just as there are places on the network of freeways where accidents are more frequent, there are places on the Earth where earthquakes are concentrated. In a way, earthquakes are the inverse of accidents. In an accident, the rapid movement of cars is interrupted by a sudden stoppage. When an earthquake occurs, the creeping motion along the boundaries of the plates is suddenly accelerated. In each case, the event is controlled by outside influences. For traffic, these may be fog, rain, or ice. In the Earth, they are the increase in pressure and deformation, and a decrease in the material strength. The critical value at which the instability in the process finally takes effect is reached through the interplay of many factors. The apparently insignificant change in some factor will initiate the process. Just as a single car accident cannot be forecast, a single earthquake cannot be predicted.

There is an additional similarity with the analogy of a traffic accident. Insurance companies can estimate the risks associated with earthquakes quite well because plate movements are predictable when averaged over 50 or 100 years. Earthquake catalogs of many countries cover several hundred years, and it is fairly easy to estimate the risks to people and infrastructure.

Earthquake Hazard and Earthquake-Resistant Buildings

Earthquakes are not solely natural disasters. More than 100 years ago, the Italian seismologist Mario Baratta wrote that people are not killed by earthquakes, but by their buildings. Technically, buildings can be made so earthquake resistant that people will not be injured or killed even in the strongest earthquake.

On a worldwide basis, the number of deaths due to earthquakes is quite large. Between 1900 and 1995, earthquakes killed more than 1.5 million people. However, large, damaging earthquakes are relatively seldom occurrences, even in active places like California, Japan, or Italy. Few inhabitants will be confronted with an earthquake that does more than expose them to a few seconds of shaking and fear. To build an earthquake-resistant structure costs more money and limits the locations available for construction. Particularly near coasts and in the mountains, the best locations for building are also the least safe. In practice, we must live with compromises between cost, use, and the willingness to accept loss of life and damage to property.

Kiyoo Mogi published an interesting statistic. On the average, the number of people killed in earthquakes annually decreased from 1900 until 1955. Since then, the number has been rising. This reversal is not influenced by fluctuations in seismic activity. The decrease was due to the introduction of building codes in regions affected by earthquakes. The rapid increase in population starting in the mid-1950s reversed this trend dramatically. Fourteen of the 25 cities with more than 10 million inhabitants are located in regions where earthquakes occur. Among these cities are Mexico City (25 million), Tokyo and its surroundings (23 million), Tehran (14 million), Jakarta (14 million), and greater Los Angeles (12 million).

There are many simple rules that can make buildings resistant to even strong ground-shaking. In 1783, Calabria was destroyed by a series of earthquakes. For the reconstruction of the city Reggio di Calabria, strict building codes, some as good as our modern ones, were enforced. The main criteria were:[4]

1. The construction style of the house should be simple and elegant. U- and L-shaped floor plans are to be avoided.
2. Houses may only have one floor and may be only 30 hand-spans high. Buildings that border public squares and broad streets may exceed this height by nine or ten hand-spans, by having a half-story on top.
3. Large balconies are forbidden. Only small, light balconies may be built. They should be attached as far from the corners of the walls as possible.
4. Iron reinforcing must be installed in all walls in all directions.
5. The interiors of houses must have a grid of well-interconnected structural timbers.
6. The addition of domes and towers is forbidden.

In the nineteenth century, seismic activity in southern Italy was relatively low. This had appalling consequences for Calabria. The building codes were ignored during new construction. In addition, the descendents who inherited a house built after the earthquakes of 1783 thought of their house as a strong foundation for more stories and large balconies. Open squares were built upon and wide streets became narrow alleys. The catastrophe was pre-programmed. On December 28, 1908, an earthquake of magnitude 7 in the Strait of Messina caused terrible damage in the cities of Reggio di Calabria and Messina and killed at least 50,000 people.

It is clear that these simple building rules are not enough for a modern society where the population density is high and structures like nuclear power plants exist. For industrial structures and high-rise buildings, earthquake safety measures must be balanced against costs and other considerations. In western industrial nations, the usual attitude toward earthquake safety is:

- Buildings and their contents must survive the effects of weak earthquakes without damage.
- For intermediate earthquakes, the building may not suffer any irreparable damage.
- For strong earthquakes, irreparable damage that can make it necessary to tear down the building is acceptable. However, the construction must prevent deaths and injury from either building parts or furnishings.

As described by Newton's Second Law, ground-shaking beneath a rigid and non-deformable structure will produce forces that are proportional to the ground acceleration and to the mass of the building. The force from an earthquake can be easily calculated. Only small, one- or two-story, strongly built buildings can be considered rigid bodies when making the calculations. For larger and especially for slender structures, the deformation that will occur must be considered. The deformation causes some sections of the building to be subjected to much stronger forces than are calculated by the simple formula of total mass × acceleration.

The elastic behavior of a building can lead to well-developed resonances that amplify the forces. If the building is complex, it can be extremely difficult to estimate such resonances by calculation or modeling. As a rule of thumb, the frequency in cycles per second of the fundamental resonance of a tall, slender structure is ten divided by the number of floors. Thus, a five-story building would have a resonance frequency of about two cycles per second. The shaking is especially strong when the resonance frequency of the building correlates with the maximum in the frequency spectrum of the exciting seismic waves. The smaller the mechanical damping of any system, the longer the resonance continues. The resonances of structures are poorly damped, usually only a few percent of the critical damping that is necessary to significantly reduce the resonance.

If buildings have such characteristics, it is understandable that it is easy to feel earthquakes in high-rise buildings even long distances from the epicenter. When earthquake waves propagate through the Earth, the higher frequencies are absorbed. At a distance of 1000 kilometers from the epicenter, the primary frequencies in the waves are 0.3 to 0.5 cycles per second. Since these are surface waves, the duration of strong ground-shaking can last one minute or more. The waves will cause high-rise buildings with 20 or more stories to resonate.

The horizontal forces applied to the bottom of a building have a particularly strong effect. The building is like an inverted pendulum, fixed at the bottom with the top free to swing. The effect intensifies when the building and the ground under it resonate at similar frequencies. This is particularly serious when loose sediment layers are lying on hard bedrock.

It is the job of the seismologist to inform the structural engineer of the ground acceleration that may be expected from an earthquake in the region. For simple structures, it is enough to give the maximum expected value—for example, 20 percent of the acceleration of gravity. For structures that require higher safety margins, the peak ground acceleration and the duration of strong ground-shaking will be used in calculations of the structural stability and integrity. This is particularly true for long pipes, such as those used in the chemical industry, or for tank storage systems.

It is not enough, however, to consider the stability of a structure. In a number of recent earthquakes, structures survived the ground-shaking. But when the ground under the structures was deformed by the earthquake waves, the structures collapsed. In effect, the ground is pulled out from under the feet of buildings or bridge supports. This can happen especially if the ground is soft and loose. In particular, water-saturated clay and sand can liquefy under the influence of vibrations.

Around the world, many cities are built in river valleys and along lake shores and seacoasts. The surface layers are primarily alluvial sand and other young, poorly compressed deposits. This has many advantages for the development of settlements. The deposits of sand and clay provide well-filtered groundwater, and wells are easy to dig. Buildings, even high-rises, are fairly easy to build on such a foundation. Problems only occur when the structures are subjected to the shaking of earthquakes. Los Angeles, Managua, Lima, Bogota, Tokyo, and Messina are some cities that are built on weak ground in regions where large earthquakes occur.

Mexico City provides a tragic example. It is located on a drained swamp. The catastrophic results of the 1985 earthquake are mainly due to the collapse of high-rise buildings that were built on piles that did not go deep enough into the ground. People reported that the high-rise buildings swayed like ships, and neighboring buildings smashed against each other until they shattered and collapsed.

Tsunamis

When the city of Messina was shaken by a strong earthquake early in the morning of December 28, 1908, many people fled their houses into the broad streets along the coast. They wanted to avoid the danger of further collapses and the fires that accompanied them. There, they experienced a unique spectacle. Just a few minutes after the shaking stopped, the sea retreated from the shore. Rocks that normally were only barely visible above the surface suddenly rose like towers from the water. The sea remained completely calm. A half-hour later, the scene was reversed. A customs official reported later that the powerful wave was as high as a house and came closer like oil, silently, without waves or foam. Hundreds of people drowned in the surge.

A *tsunami*, a compound Japanese word meaning "harbor wave" (*tsu* = harbor and *nami* = wave), is a series of flood waves that are generated by a sudden vertical displacement of the ocean floor. Earthquakes are the most common cause of major tsunamis. Other sources are submarine volcanic eruptions and landslides. The location of the Messina earthquake was in the narrows between Calabria and Sicily, and the source process caused the seafloor to sink by 1 meter. The largest tsunami ever recorded was generated in May 1960 by an earthquake that occurred off the coast of Chile. The tsunami crossed the Pacific Ocean to the Japanese coast in 22 hours. Ships on the ocean did not notice the tsunami. Because the waves were about 1 meter high and had a period of 10 minutes, the accelerations were much too small to be felt at sea. That changed when the waves reached shallow water near the coast. The propagation velocity of a tsunami depends on the depth of the water. In the deep ocean, it is about 700 kilometers per hour, but it decreases rapidly when the water gets shallower. Because energy must be conserved, the wavefront then becomes higher. This is why the tsunami generated by the Chile earthquake produced waves over 20 meters high more than 15,000 kilometers away in Japan and killed 200 people.

Physically, tsunamis are surface waves in the water. Their propagation velocity v is given by the relationship

$$v = \sqrt{g \times d}$$

where g is 9.81 meters per second squared, the acceleration due to gravity, and d is the depth of the water. Because gravity appears in the equation for tsunamis' propagation, they are also called *gravity waves.* Since tsunamis propagate much more slowly than seismic waves, it has been possible to develop a warning system that is particularly effective in the Pacific Ocean basin. A few minutes after an earthquake occurs, the location, strength, and type are calculated with the help of seismogram data. The type of earthquake is used to determine the possible vertical displacement of the seafloor. If a tsunami is expected, endangered coastal cities are warned to expect a wave of a certain size at a certain time. Despite the successful work of the Pacific Tsunami Warning Center in Honolulu, Hawaii, issuing warnings is not without problems. A strong earthquake in Alaska on Good Friday, 1964, caused a tsunami that propagated throughout the entire Pacific Ocean. When people in California heard the warning on the radio, thousands of them streamed to the coast to see the natural spectacle, and twelve people died.

1 Alexander von Humboldt, *Kosmos: Entwurf einer physischen Weltbeschreibung*, Vol. 1 (Stuttgart, Germany: Verlag J. G. Cotta, 1845).

2 Rudolf Hoernes, *Erdbebenkunde* (Leipzig, Germany: Verlag Veit und Comp, 1893).

3 Harry O. Wood and Frank Neumann, "Modified Mercalli Intensity Scale of 1931," *Bulletin of the Seismological Society of America* 21 (1931): 277–283.

4 Albin Belar, *Die Erdbebenwarte* 8 (1909): 59.

chapter 3

Volcanoes

Like earthquakes, volcanoes have both fascinated and terrified us for thousands of years. Today, although much better understood, volcanoes are still shrouded in mystery. This chapter gives an introduction to what volcanologists have discovered about volcanoes, and what questions and challenges still remain. We'll begin by exploring the historical development of volcanology. Next, we'll explain how and where volcanoes form as well as what mechanisms and circumstances drive and define the eruptions. Specific volcanoes in North America will be discussed as well as the still mysterious, so-called *hot spots* and the evidence of volcanic activity in our Solar System. In conclusion, we'll present the challenges faced by those who monitor these explosive and often deadly landforms.

Fundamental Questions

> If we look back at the ideas which have been proposed as explanations of volcanic phenomena by various researchers from ancient times until the present, we perceive considerable overall progress, despite occasional steps backwards, corresponding to the growth in the sciences. Thus, we may trust that by continuing with our patient work and careful thinking, we will slowly approach our goal of improved knowledge, however distant it may be at this time. Collection of new, reliable facts must be the most important goal of future studies, as only such facts can someday provide a strong foundation for more satisfactory theories than have yet been conceived.

These are the closing sentences of the 1927 book *Volcanology*, written by one of the grand old men of volcanology, German professor Karl Sapper.[1] His comments can still be applied to volcano studies at the beginning of the twenty-first century. Great progress has been made in all areas of volcano research in recent decades. This is especially true of the techniques for experimental observation and the methods for quantitative analysis of parameters from geophysical, geochemical, and petrological measurements. These modern methods allow the volcanologist new insights into volcanic processes. They include the use of new seismographs, real-time transmission of data from remote sites to observatories for immediate analysis, and satellite monitoring of volcanoes for thermal, geodetic, or geochemical changes. Today, we can agree with Sapper that recent years have brought new and important knowledge. However, this can be seen in the development of more precise questions rather than in well-grounded answers.

For example, lava fountains, which last for hours or days and are several hundred meters high, pose more puzzles for the modern volcanologist who thinks in terms of the principles of physics and fluid dynamics than they did for his colleagues from earlier times. Volcanologists want to know which changes in the eruptive behavior of a volcano are important—why and when does a volcano make the transition from inactive to active? The answers to these basic questions are as open today as they were in earlier times.

Few volcanoes in the world are as well studied as Vesuvius near Naples, Italy. The world's first institute for the physics of the Earth was founded on its flanks in 1857, and the first seismograph was set up there. More than one thousand scientific articles have been written about the structure and volcanic activity of the mountain. Despite the high level of knowledge, one fundamental, and for the surrounding countryside, very important, question cannot even begin to be answered. Why is the great crater of Vesuvius quiet now, after a period of nearly uninterrupted activity between 1631 and 1944? Vesuvius's activity began more than 15,000 years ago. If only for the reason that geological processes tend to continue over long time spans, the volcanic activity of the mountain is not likely to have ended for all time in 1944. Yet, even with all the knowledge of modern volcanology and high-quality measuring methods, we do not have even marginally satisfactory answers to questions about its activity. Will the next eruption come in one week or in one thousand years? What will the transition from quiescence to activity look like? Will this phase last one hour, or a week, or longer? Which visible or instrumentally observable precursors will precede the new activity? Earth scientists attack these questions without knowing whether or when they will ever be answered.

In the natural history of ancient times, volcanoes were thought to be unimportant side effects of earthquakes. The ancient Greeks had little experience with them, as, in historic times, eruptive activity on the Greek mainland occured either sparsely or not at all. Other important volcanoes, such as Vesuvius and the volcanic islands of the Aegean Sea, were relatively quiet during this period. The eruptive activity of the Aeolian Islands was noted in documents, but the islands were off the beaten paths of the time and so attracted little attention to the phenomenon of volcanism. Only Mount Etna on Sicily came to their notice. Up until the late Middle Ages, all fire-emitting mountains worldwide were thus called "Etna" or, alternatively, "Hiera," for the fire mountain on the island of Hiera in the Tyrrhenian Sea, north of Sicily. Only later were they named *volcano*, following the Roman name "Vulcano" for the same island. Aristotle believed that air trapped under pressure in the Earth was the cause of earthquakes. He explained the fiery manifestations of volcanoes as the ignition of the compressed air. From the Stoic Posidonius, we have detailed descriptions of volcanic activity in the Mediterranean. He viewed combustible materials as a cause and listed sulfur and tar (asphalt or pitch) as providing heat for volcanoes. Later, in the Middle Ages, the ideas from the Greeks and the Romans were combined with Christian beliefs from the Bible. Volcanoes became the gates to the hell fires of pitch and sulfur. In the name of religion, fantastically speculative ideas were proposed.

The end of the eighteenth century brought many great discoveries in physics and chemistry. In their enthusiasm, researchers well known for their work in these areas made suggestions about the nature of volcanic heat and possible sources for it. Supported by the French physicist Louis Joseph Gay-Lussac, the English chemist Humphry Davy proposed that volcanism is based on heat-producing reactions caused when water comes into contact with alkali metals and other unoxidized elements. Another explanation from this era related volcanoes to the still rather mysterious phenomenon of electricity. For example, static electricity in the Earth was thought to ignite hydrogen to produce eruptions.

The beginning of modern volcanology dates from the work of Sir William Hamilton. At the end of the eighteenth century, Hamilton was the English ambassador to the king's court in Naples. He was fascinated by Vesuvius, which was extremely active at the time, and became an enthusiastic volcanologist. In contrast to other scientists of the time, he realized that reliable explanations for the source and nature of volcanism could not be made, as he said, from the comfort of a chair. On the contrary, they should be based on the interpretation of extensive field observations. He was the first to have volcanic rocks and gases analyzed chemically, and he distinguished between different types of flow in lava streams. He also demonstrated that volcanic mountains develop as accumulation cones. Based on many observations, he came to the important conclusion that the source of volcanic fire lies at great depth. He also thought that volcanism is a remnant of processes that occurred during the creation of the planet. The great German author and naturalist Johann Wolfgang von Goethe met with Hamilton several times during his trips to Italy. Despite Hamilton's persuasive arguments, von Goethe refused to change his opinion that volcanism is based on "fires in the earth as a consequence of wide-spread coal deposits."[2]

Hamilton's extensive field investigations of Vesuvius and other volcanoes, often dangerous, were the key to his success. Many of the physicists and chemists of the day preferred their laboratories to the difficulty and exertion of climbing volcanoes and making observations. When their speculations about volcanoes met with little success, they quickly lost interest. Volcanology became the domain of geographers and geologists. The German chemist Robert Bunsen was one of the few exceptions. At the end of the nineteenth century, he completed work on the eruption process of geysers that is still accepted today.

Volcanoes represent a window into the otherwise inaccessible depths of the Earth. Hamilton prepared the path for the realization that we can draw conclusions about the Earth's interior on the basis of extensive, systematic, and scientifically sound observations and data from volcanoes. To explain how volcanoes work, scientists must try to construct a bridge from their observations and data to scientific laws. The data represent and describe only a portion of the phenomena occurring at volcanoes. Technical and financial limitations necessitate compromises in the measurements. In addition, the observation time is limited to a small fraction of the life span of a volcano. This is why different—even mutually exclusive—models of how a volcano works can arise. Combining observations of volcanoes with new knowledge (for example from fluid dynamics or the behavior of material at high temperature and pressure) can limit the choice of possible models. However, even here a type of uncertainty principle is valid: the more precise the claim, the more questionable is its validity.

Melting Processes in the Earth's Mantle

Volcanoes are perhaps best known for their dramatic and dynamic explosions that throw smoke, ash, and glowing lava into the air. But where does it all come from? We will begin our discussion of volcanic dynamics by exploring the origins of the most showy of the ejecta: lava.

Magma Production at Divergent Plate Boundaries

Until the end of the nineteenth century, the Earth was assumed to have a 20- to 50-kilometer-thick crust everywhere, covering an ocean of melted rock, or *magma.* The first studies of the propagation of earthquake waves proved this was wrong. Aside from the metallic core at a depth of 2900 kilometers, the shear strength of the Earth corresponds approximately to that of steel. Thus, it cannot be considered to be liquid. At first glance, this may seem surprising because many of the young rocks on the Earth's surface have apparently formed by solidifying from melted material. And from lava flows at volcanoes, we know that melted rock must exist inside the Earth.

It was not until the middle of the twentieth century that measurements of seismic waves became precise enough to detect a small decrease in the propagation velocity of shear waves in the upper mantle and a small increase in their absorption. By comparing this with results from high-pressure laboratory experiments on rocks, scientists concluded that the temperature between depths of 100 and 200 kilometers is very close to the melting temperature of mantle rocks. Under such conditions, it is likely that *partial melting* occurs. This means that some, but not all, components of the heterogeneous mantle rocks are melted.

No plausible explanation for the development of partial melting in the Earth was found until the theory of plate tectonics was developed. This theory postulates the existence of convection currents in the mantle that drive the plate motion. Let us examine what happens to the convection currents at the mid-ocean ridges.

We will select a small volume of mantle rock from a depth of 200 kilometers. As the sample rises in the convection current, the surrounding pressure drops. Heat loss from the sample by conduction is negligible at a flow rate of several millimeters per year, which is rapid in comparison with other processes deep in the Earth. We call a pressure drop without heat transfer adiabatic decompression. Pressure experiments have been done on materials like those found in the mantle. They show that phase transitions, solid-to-liquid or liquid-to-solid, occur at lower temperatures when the pressure drops. The mantle rock is heterogeneous; it is composed of many different mineral components. At a given pressure, melting and freezing do not occur at the same temperature. The melting and freezing curves are separated. At temperatures and pressures below the freezing curve, all mineral components are solid. Above the melting curve, all are liquid. Between the two curves, the rock has a consistency somewhat like a water-saturated sponge. In the solid matrix of the granular skeleton are liquid droplets of materials that have already melted.

Peridotite is a rock typically found in the upper mantle. Let this rock rise from a temperature environment of, say, 1300 °C. When the lithostatic pressure reaches 30 kilobars, at a depth of about 100 kilometers, the material passes the freezing curve and reaches the zone of partial melting. As it continues to rise, more and more rock components melt. Latent heat is necessary for the melting to proceed. This heat is drawn from the sample itself, causing the temperature to decrease. Whereas the temperature gradient in the frozen phase is about 0.5 °C per kilometer, it rises in the zone of partial melting to a value that is ten times higher. This change in the temperature gradient stabilizes the material in the lithosphere. Otherwise, the Earth's surface along the mid-oceanic ridges would be covered by giant seas of magma.

At the center of ocean ridges, the convection currents come closest to the surface, to a depth of about 6 kilometers. The current can no longer rise, but is diverted to flow horizontally. It cools as the distance from the ridge axis increases, and the lithospheric plates become thicker as some of the melt adheres to them. As the current cools, the percentage of partial melt in the rock decreases. This percentage depends on the temperature and pressure in the rock. At the ridge axis, the melt makes up 20 percent of the rock. At a distance of a few hundred kilometers, it has decreased to only a few percent. The chemistry of the rock changes correspondingly. At the ocean ridge, the rocks are mainly tholeiitic basalts, while farther away they are mostly alkali basalts.

The melted droplets in the matrix of solid rock are the beginning of the chain that ends as volcanism at the surface of the Earth. The existence of the droplets is not enough, however. Next, the droplets must separate from the solid material and collect as a fluid. For this to occur, the microscopic spaces between the grains of the minerals containing droplets must be connected by capillary-like channels.

The density of each component in the melt is lower than its corresponding solid. According to Archimedes's principle, the droplets of melt in the network of capillaries are buoyant because their density is lower. That is, they experience an upward force. The buoyant force of the droplets exerts a force on the walls of the solid rock surrounding them, causing plastic deformation that pushes the walls apart. As the process continues, the capillaries grow larger and eventually become conduits for the collected melt, which is magma. The flowing magma can break pieces out of the solid rock surrounding it and transport them to the surface as though they were in an elevator. These foreign bodies are called *xenoliths.* They are our only way to collect relatively unaltered rocks from the Earth's mantle. Xenoliths have also been called "meteorites from the interior of the Earth." They can be as large as a football, which means that the conduits in the mantle are larger than that in diameter.

Only a small portion of the magma arrives at the surface. As it rises, the density of the rock surrounding it decreases, and therefore the magma's buoyancy also decreases. There would

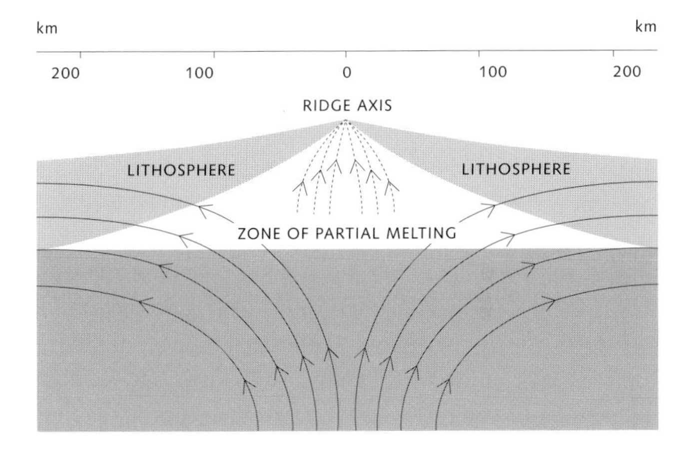

Cross-section perpendicular to a mid-ocean ridge. The lines represent the flow paths of ductile, flowing rock rising from the mantle. The flow diverges at the ridge axis, moving symmetrically to both sides. As the distance from the axis increases, the flow cools and the lithosphere becomes thicker. Beneath the axis a zone of partial melting develops. The melt flows upward toward the ridge axis and causes the volcanism there.

(After a drawing by Steven Sparks.)

probably be no volcanic activity at the Earth's surface if no force other than buoyancy were involved. However, there are also tectonic forces present. If horizontal compression is present in addition to the lithostatic pressure, then mountains are raised. If the added forces are dilatational, then extension structures and rifts develop. Working against a lithostatic pressure that is reduced due to horizontally-oriented tensional or dilatational forces, the buoyancy of the magma is strong enough to open small vertical channels into larger cracks and fissures. Only then can the magma create paths toward the surface. Thus, there must be a conjunction of favorable factors for volcanism to appear at the Earth's surface. Eruptive volcanism is a relatively rare occurrence. It is not easy for an interested tourist to experience eruptive volcanic phenomena in the form of lava flows or large ash clouds, even by travelling to well-known, active volcanic areas.

Magma Development at Convergent Plate Boundaries

About ten percent of the today's active volcanism occurs at mid-ocean ridges. Eighty percent takes place in subduction zones at convergent plate boundaries. The Ring of Fire around the Pacific Ocean is typical subduction volcanism. Melting caused by the decompression of hot, rising mantle rocks seems to be a plausible explanation for the hot mantle material along the mid-ocean ridges. However, it is more difficult to explain the presence of volcanism at subduction zones. The lithospheric plates descend precisely because they cool and become denser. The water taken up by the rock of the plate during its trip from the mid-ocean ridge, where it was formed, to the subduction zone plays an important role in volcanism. Near the ridges, the sheets of basalt are hot and have many deep fissures. Through hydrothermal processes at depths up to several kilometers and at temperatures higher than 300 °C, basalt rocks are altered into hydrated minerals such as amphibolite and serpentinite.

During transport from the ridge to the subduction zone, the basalts are usually covered by a thick layer of marine sediments and water-containing rocks. As the lithospheric plate descends into the hot mantle, it heats slowly, by conduction. As it descends, the material of the plate is subjected to complex chemical and mineralogical alterations. Exactly what these alterations are and how they occur are the subjects of a multitude of speculative assumptions. As the temperature increases, the rocks dehydrate and water molecules and hydroxyl groups collect between the silicates. This decreases their chemical activity and reduces the melting point of the minerals by several hundred degrees. As the lithostatic pressure increases around the descending plate, some of the silicates dissolve in the water, reducing the melting point even further.

Here again, the less dense components of the fluids that develop experience buoyancy due to Archimedes's principle. They separate from the subducting plate and react with rocks of the upper mantle on their way towards the surface. The belt of volcanoes above the subducting plate is rarely more than a few hundred kilometers wide, and is therefore narrow compared to its length. The development of melt in the subducting plate takes place only between depths of 80 and at most 200 kilome-

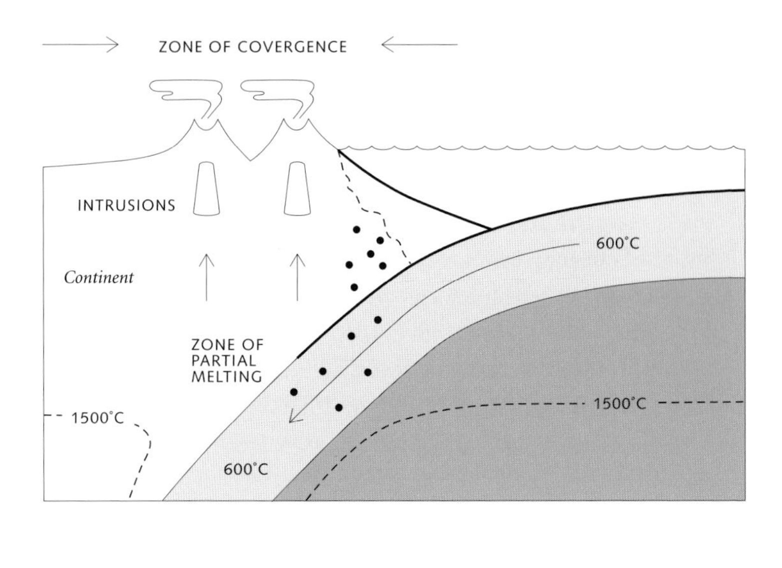

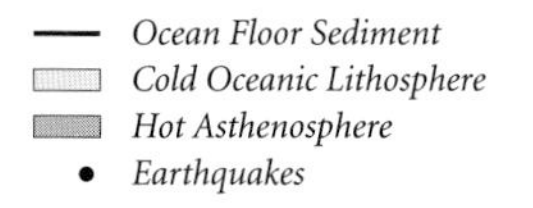

Melting processes and the development of volcanoes at a subduction zone.

ters. It is confined to the temperatures and pressures corresponding to this depth range. Earthquakes are also more likely to occur at these depths in the subduction zone.

The volcanism associated with convergent plate boundaries and subduction is complex. It has many different types of eruptive activity, diverse output products, and produces a great variety of landscapes. We can distinguish between two cases: An oceanic plate may be subducted beneath either an oceanic plate or a continental plate. On the overriding plate, we observe a sequence of landforms. As we approach the overriding plate, we encounter an accretionary wedge of folded sediments and pieces (*ophiolites*) scraped off the top of the descending oceanic plate. The accretionary wedge is usually uplifted and is followed by the *forearc basin*. This basin collects sediments eroded from elevated parts of the accretionary wedge, as well as from high terrain further along in the direction of subduction. At depths of 80 to 200 kilometers, partial melting occurs in the subducting plate. If the overriding plate is oceanic, the rising magma produces an *island arc*. Indonesia, the Aleutians, the Mariana Islands, and the islands of the Caribbean are examples of island arcs. When the overriding plate is continental, a broad belt of volcanoes develops along the edge of the continent. The Andes on the western edge of South America and the Cascades in the Pacific Northwest of the U.S. are volcanic belts. In some cases, *backarc basins* are observed behind the volcanic belts. They usually involve thinning of the crust, often with subsidence or rifting and more volcanism. The causes are the subject of controversy. Backarc basins can develop into marginal seas, like the Sea of Japan or the Java Sea. Some scientists also consider the Aegean Sea to be a backarc basin.

The chemistry of volcanic products produced in subduction zones depends mainly on the type and thickness of the rock matrix that the melt must penetrate on its way to the surface. Along geologically young island arcs like the Mariana or Tonga Islands, the melt encounters neither continental crust nor thick layers of sediments. The volcanic products resemble the silica-poor basalts of the mid-ocean ridges. At the volcanic belts on the edges of the continents, the melt comes in contact with silica-rich layers of crustal rocks up to 70 kilometers thick. The buoyancy of the melt in these low-density crustal rocks decreases because the difference in density is less. It is likely that most of the melt from great depths no longer reaches the surface.

Igneous Rocks

Igneous rocks are crystallization products that develop when magma, the melted rock that has risen from great depths, cools. Magma is usually a heterogeneous mixture of silica-rich materials containing dissolved gases and crystals in addition to the melt. If the magma does not reach the Earth's surface but slowly cools at some depth, it is called an *intrusion* and the results are medium- to large-grained plutonic rocks. Granite is such an intrusive rock. As magma rises, the lithostatic pressure decreases and it begins to degas (lose gas). The degassed magma that flows out at the surface is called *lava*, from the Latin word *labes*, meaning fall or landslide. In southern Italy this word is used for flash floods as well as for streams of molten rock from volcanoes. *Volcanic rocks* develop from the cooling lava. They may form from lava that has flowed quietly from a volcanic vent or that has been ejected explosively. During explosive volcanic activity, liquid or solid lava fragments ranging from the size of dust particles to blocks weighing several tons can be thrown into the atmosphere. These are called *pyroclastic rocks.* Volcanic ash can fall to the Earth, producing immensely thick layers of sediments that can solidify to *tuff.* Uncemented pyroclastic deposits are called *tephra.*

Description of Igneous Rocks

Igneous rocks have little in common with the rocks from which they began. The original partially melted material, or *primary magma*, starts its existence as a derivative of the asthenosphere or the subducting plate. As it rises, it reacts thermally and chemically with the rocks around it. Under some circumstances, the change in the primary magma is caused simply by an exchange of heat. This happens when heat from a denser substance melts a less dense material with a lower melting temperature. The dense mineral then remains behind and only the newly melted material rises.

The more time a packet of melt needs for its ascent, the more changes it can undergo. If the melt cools at some point during its rise to the surface, then the partial melting process is reversed. The melt experiences fractionating crystallization: the mineral components with the highest melting temperature crystallize out. At some point it is possible that these crystals melt again. From these possibilities, it is clear that the term "magma" is very general.

The investigation and classification of volcanic rocks and their derivatives is the task of *petrology*, the study of rocks. Relatively simple mixtures of magma can develop into any number of rock combinations, depending on the pressures and temperatures they experience during their ascent, their upward velocity, and the chemistry of the materials they pass through. At first, rocks were named and classified according to where they were found. Today rock names depend on the rock's composition and the methods used for its determination. As the methods become more discriminating, rocks of different types can be more readily distinguished. However, historically developed names and classes are difficult to eliminate. Additionally, the composition of rocks represents a continuum of minerals without clear boundaries. Thus, the confusion that has developed is difficult for an outsider to comprehend. Today there are more than a thousand different names for volcanic rocks.

Originally, volcanic rocks were differentiated on the bases of characteristics that could be seen with the naked eye: color, grain size, and the types of crystals in the rock. Dark lavas were called basalts. They were subdivided according to the visible crystals they contained, for example olivine basalts or feldspar basalts. If no crystals were visible, they were called *aphanites. Phonolites* were light-colored lava that rang when hit with a hammer. From the very beginning, petrologists recognized that the amount of quartz (silica, SiO_2) in the rock was an important distinguishing characteristic. Since they considered the silicates to be the result of precipitation from silicic acid, petrologists indicated the amount of SiO_2 by calling the rocks *acidic* or *basic.* Today we distinguish them by calling rocks with high quartz content *felsic* and those with little quartz *mafic.* This can be determined from the ratio of the areas of light and dark grains in a thin section of the rock under a polarizing microscope.

There is often a close relationship between the texture of an igneous rock and its origin. From the grain sizes of a solidified volcanic rock, we can derive information about the viscosity and crystallization of the magma during transport, as well as the rate at which it cooled. In a low-viscosity magma that cools slowly, the crystallization nuclei have time to develop into large crystals. Large-grained rock texture is typical when the solidification has occurred at great depths. The phenocrysts that can be seen as large crystals in volcanic rocks often develop while the melt is cooling slowly. The cooling process may be forcefully interrupted by an eruption. The uncrystallized portions of the magma then crystallize rapidly, producing a fine-grained texture. Sometimes, when highly viscous lava containing a large amount of silica cools rapidly, no crystals develop. The result is volcanic glass, or *obsidian.* Thus, the rock texture, determined by examination under a microscope, gives clues to the origins of rocks and the geological classes to which they belong.

The modern classification of volcanic rocks usually relies on mineral composition and chemical characteristics. Rocks are composed of one or more minerals. Feldspars play a prominent part in magmatic rocks and in the composition of the Earth's crust. Chemically, feldspars are silicates with relatively simple structures. They have silicon as their main constituent and, in varying amounts and in varying structures, sodium, potassium, calcium, and aluminum. Often magmatic rocks are classified by plotting the percentages of quartz, alkali feldspar, plagioclase feldspar, and feltspathoids in a double triangle diagram. The corners of the double triangle represent the four mineral classes.

In other classification methods, the weight percent of alkalis (Na_2O and K_2O) in the rock are plotted on a diagram as a function of the SiO_2 content. Rocks of different families are separated in the diagram. The different groups are designated by the type of source magma as granitic, dioritic, and so on. We can differentiate three groups of rocks by their chemistry. They are the calc-alkaline group, the alkaline group with mostly sodium, and the alkaline group with mostly potassium. The most important rocks in the calc-alkaline group, in order of descending silica content, are rhyolite, rhyodacite, dacite, andesite, tholeiite, and picrite. There is an important correlation between the various rock groups and the tectonic setting in which they are found. For example, silica-poor volcanic rocks, to which the large class of tholeiites belongs, mainly occur at mid-ocean ridges. Silica-rich rocks are found in subduction zones.

Eruption Mechanisms

After a long period of quiescence, a small white cloud of steam was observed at the northeast crater of Mount Etna in Sicily at about noon on September 24, 1986. This was not unusual. At least one of the four summit craters of Etna is always active with steam or ash emissions. Large explosions with big volcanic bombs are relatively rare. This date, however, entered the ledger of Etna's eruptive history. In less than an hour, the diffuse cloud had developed into tightly bundled, hissing jets of steam that climbed several hundred meters into the sky. At about 2 P.M., the color of the volcanic emissions changed from white to black. The magma in the crater began to boil and the degassing became very strong. Droplets from the pool of lava were caught by the thermal air currents and blown as cooled, solidified, black ash from the crater mouth.

Often, this type of phenomenon signifies the maximum activity and it quickly dies down again, disappointing volcanologists and tourists. On this day, it was not to be. At about 3 P.M., the explosive activity not only became stronger, but the belching became rhythmic, occurring every second or so. As a result, a 1-kilometer high eruption cloud developed, shaped like a giant cauliflower. The falling pyroclastic material increased in size. Even 6 kilometers from the crater, the lava fragments, called *lapilli*, were the size of cherries and walnuts. Impressed by the rare spectacle and assuming that the eruption had reached its pinnacle, a well-known photographer of Etna shot all his rolls of film. Against all the expectations of the experts, the eruption had still not reached its climax. Shortly after 6 P.M., the cloud began to glow. This was quickly followed by a lava column more than 1000 meters high above the crater. Dark spots were observed in the fountain. They were blocks weighing several tons that had been ripped from the walls of the crater, found later distributed around Etna's summit area. The trajectories of some of the blocks could be followed with the naked eye. They had flight times of up to a minute, indicating initial ballistic velocities of more than 100 meters per second.

There are, however, much stronger eruptions. During the 1991 eruptive activity of Mount Pinatubo in the Philippines, observers estimated that 1 million cubic meters per second were thrown from the volcano's vent at velocities up to 600 meters per second.

Where does the driving force come from? Buoyancy forces responsible for the rise of magma in the mantle and crust are several orders of magnitude too small to explain the high exit velocities of volcanic material. For obvious reasons, observations of the dynamics of eruptions must be made from a safe distance. Attempts to move gas analysis equipment into active volcanic craters using high-tech, relatively expensive robots have failed. For example, at Etna, radio-controlled model airplanes were to extract samples from the middle of the eruption cloud, as the volcanic gases at the edge are contaminated by the atmosphere. Although the European model airplane champion was hired to control the planes, the experiment had to be cancelled because too many planes crashed in the turbulence of the emission cloud.

Seismic measurements allow an indirect look at the flow of magma and the pressure changes occurring during an eruption. Flowing magma produces noise similar to the murmur and gurgle of water flowing in a pipe. Fluctuations in the flow velocity of the magma are associated with changes in pressure. The pressure variations act on the walls of the conduit and produce seismic waves that propagate through the solid Earth. The amplitude of these waves decreases slowly with distance from the source. The waves can therefore be recorded at a safe distance from the eruption. From the tone of water-induced noises in a pipe, we can tell if the flow is fast or slow. Similarly, we can estimate flow characteristics from the frequency spectra of seismic recordings of magma movement. Sometimes a hard, often rhythmic, hammering noise is superimposed on the soft murmur of flow in a water pipe. We call this *water hammer.* Several years ago, waterlines in the water supply system of Stuttgart, Germany, burst due to high pressure that could not be explained. The puzzle was finally solved: Air bubbles in the water caused the flow to become unstable at places where a waterline bent, was constricted, or branched. The instability caused hydraulic pressure pulses. The pressure increased to several times the static level for a brief period during a pulse, causing the pipes to burst.

Many investigations and hypotheses suggest that dynamic pressure variations related to flow instabilities are the driving force for the explosive ejection of magma. The solubility of gases in the magma decreases as the magma rises and the lithostatic pressure decreases. If the vapor pressure exceeds the lithostatic pressure, bubbles develop. Water vapor and carbon dioxide are the main gases. Under normal conditions with no particularly high activity, Etna emits 50,000 metric tons of carbon dioxide daily, and many times that of water vapor.

The flow behavior of bubble-enriched magma is very different from that of magma without bubbles. Its dynamic behavior is much more complex. In general, it is typical for high-intensity pressure pulses to occur when the flow of a bubbly fluid exceeds a certain velocity. One well-known flow instability that produces a strong pressure pulse occurs when the flow velocity exceeds the speed of sound in the medium. In air this is known as a sonic boom. The speed of sound in a fluid depends on its density and compressibility. Most liquids cannot be com-

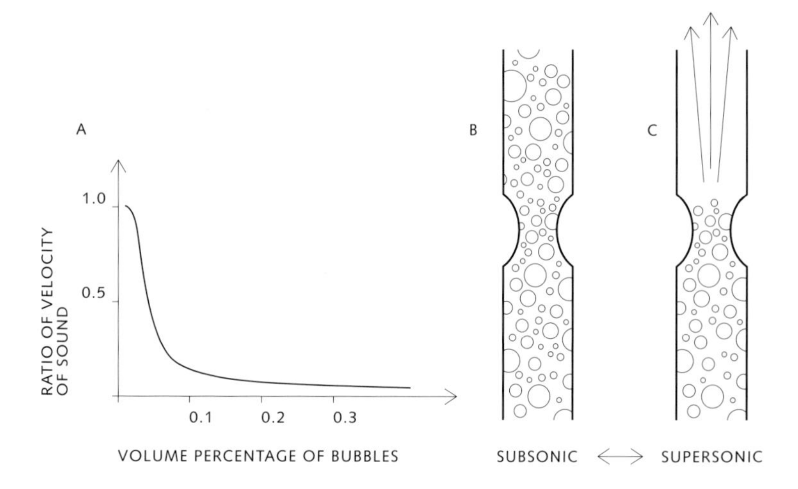

During volcanic eruptions we often observe a rhythmic output of volcanic products from the crater. It is likely that this process, which can continue for hours or days apparently without changing, is stabilized by a feedback cycle. The mechanism of this cycle may be as follows:

When gas bubbles are added to a liquid (the melt), the speed of sound in the two-phase fluid drops sharply compared to that of melt with no bubbles (A).

In a basalt melt with several percent by volume of gas, the speed of sound is 10 to 20 meters per second. Solidified and eroded conduits often have constrictions. Due to mass conservation, the flow velocity of the magma must increase in narrow places. This reduces the pressure and increases the volume of bubbles (B).

The increase in the percentage of bubbles means that the speed of sound decreases again. This continues until the speed of sound is the same as the flow velocity. When that happens, a pressure pulse, or sonic boom, develops. The pressure pulse pushes the melt and gas toward the free surface (C), reducing the volume percentage of the bubbles and thereby returning the system to its original subsonic conditions. Such cycles are a possible explanation for the driving forces and the pulsations of the eruption products.

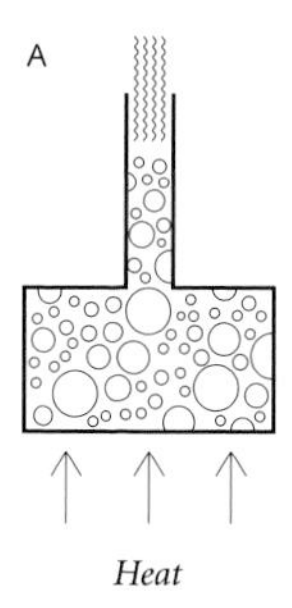

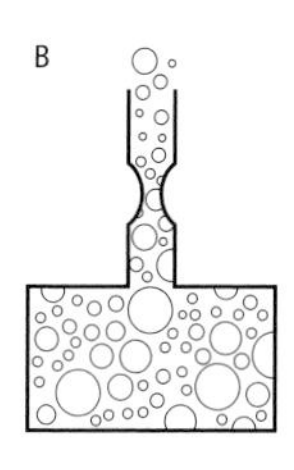

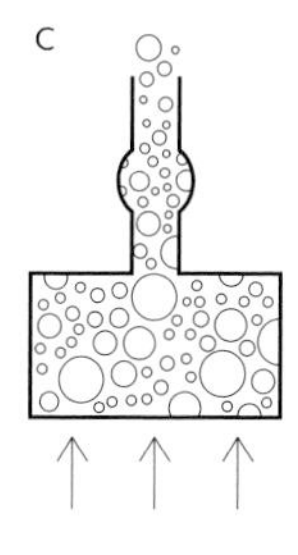

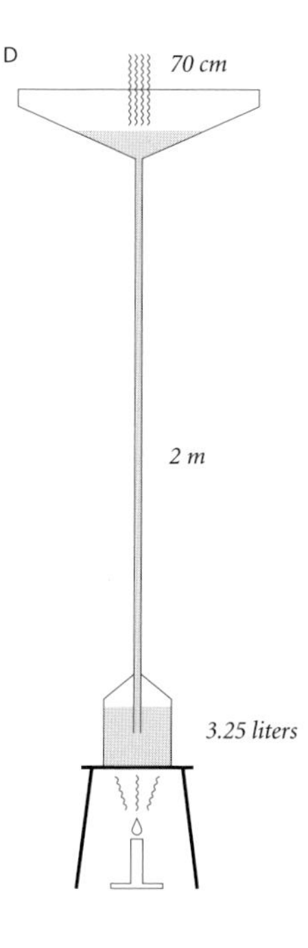

Not all aspects of the dynamic processes occurring during volcanic eruptions can be reproduced in laboratory models. One important reason is that when the dimensions are reduced for the model, volume and surface forces decrease differently. Nonetheless, some typical sequences of eruptive processes can be simulated qualitatively in small dimensions.

A standpipe or hose is mounted on a container filled with water (A).

If the bottom of the container is heated, steam bubbles quietly leave the top of the pipe when the water boils. If, however, the cross-section of the pipe is reduced at one place (B), a rhythmic, explosive discharge of steam develops in which water is carried along with the steam. The process has similarities with "Strombolian activity."

A widening in the pipe ("magma chamber") has a similar effect (C).

Experiments of this sort were carried out by the end of the nineteenth century: (D) shows Andreäs's apparatus for experimentally demonstrating geyser activity.

(Figure (D) after A. Sieberg: Der Erdball. *Esslingen & Munich, Germany: J. F. Schreiber, 1908.)*

pressed very much. However, with the addition of only a few percent of bubbles by volume, the compressibility of the mixture depends mainly on the compressibility of the gas, while the density depends mainly on the liquid. The result is a drastic reduction in the speed of sound. In bubble-free magma, the speed of sound is about 2000 meters per second. With 1 percent gas, it drops to about 20 meters per second. As the fraction of gas increases, the speed of sound falls to levels corresponding to the flow velocity of the magma. When this happens, a pressure pulse develops that pumps the magma upward.

Volcanic eruptions are often compared to opening a bottle of champagne. However, the analogy is not accurate. When the cork leaves the champagne bottle, a rapidly decreasing stream of liquid and gas shoots from the bottle. In contrast, the activity in volcanoes can last for hours or days without decreasing in intensity. Apparently, feedback mechanisms control and stabilize the dynamics of the eruption. With the help of seismic measurements, we can follow the change in the mean flow velocity of the magma as the activity increases. The time functions are very similar to the turn-on and turn-off transients of oscillators. Perhaps the network of magma-containing channels and reservoirs in the volcano can be compared to the structure of a fluid oscillator—a hydraulic construction element that enhances dynamic flow processes. If this analogy to feedback oscillators is valid, then the following concept could explain the difference between quiescence and activity in a volcano: The feedback necessary to start an eruption needs more than just a reserve of potential heat energy. Other factors such as the material characteristics and geometric dimensions of the conduits

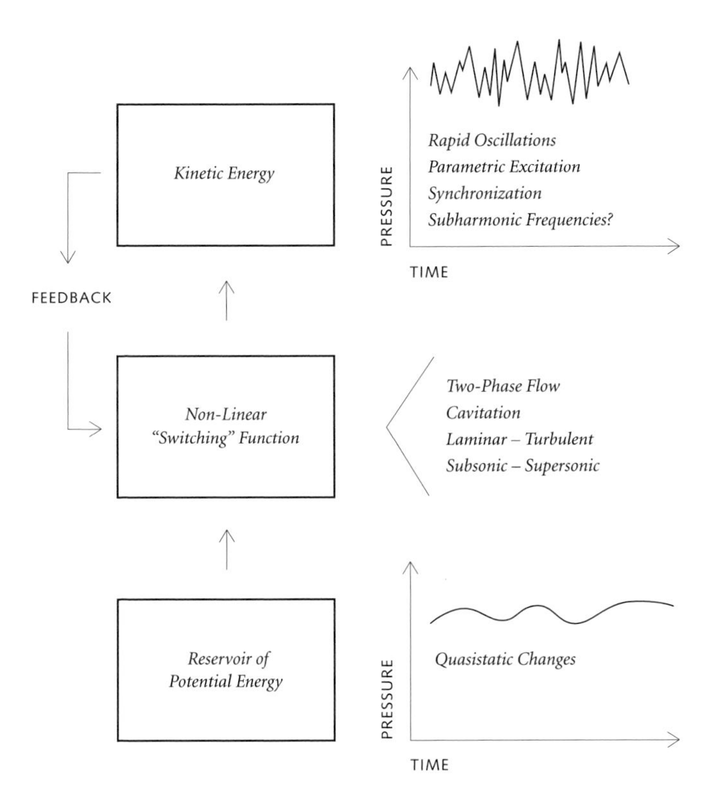

The mechanism in an active volcano is similar to a steam engine. Both produce kinetic energy from heat derived from a reservoir of potential energy. Slow, quasi-static pressure changes are transformed into a rapid sequence of pressure pulses with the aid of a switch controlled by a feedback mechanism. In the steam engine, the switching function is provided by the mechanical slide valve; in the volcano the switches are elements described by non-linear terms in the flow equations. Among these elements are bubble growth and collapse (cavitation) as well as transitions between laminar and turbulent flow or subsonic and supersonic flow. The sequence of rapid pressure changes has characteristics typical of non-linear oscillators.

determine if self-sustaining dynamics can be achieved. If certain geometries of the structure fit, a small disturbance in one of the parameters controlling the process is enough to initiate the transition from non-erupting to erupting. A volcano is similar to a steam engine. With the help of transformation elements that act non-linearly, potential energy in the form of heat is transformed into kinetic energy—that is, energy of motion. In the steam engine, the slide valve is the non-linear element. In the volcano it may be gas and steam bubbles.

The Explosive Effects of Volcanoes

The strength of an earthquake can be described easily and precisely enough using its magnitude and the distribution of the intensity of shaking. Because of the many different forms of volcanic activity, it is difficult to define useful parameters for describing the strength of an explosion. Such parameters would be useful not only for classifying volcanic eruptions in catalogs but also for quantifying the strength of an eruption for statistical investigations of the influence of volcanic eruptions on the weather. Although it would be almost presumptuous to assign a single number to a concept as broadly defined and as subjective as the strength of an eruption, the *Volcanic Explosivity Index (VEI)* is used as a compromise. This parameter was developed on the basis of proposals by the volcanologists Karl Sapper (1920) and Alfred Rittmann (1935). The VEI is determined using five parameters:

- the volume of the eruption products;
- the production rate (volume per unit time), which is determined from the exit velocity of the material and the height of the eruption column;
- the range of the eruption, determined from the height of the eruption column;
- the strength of individual explosions;
- the potential for destruction.

To achieve similarity with earthquake magnitudes, the VEI is given numbers between 0 and 8. An increase of one unit should correspond to an increase in the parameters by a factor of 10. In comparison with historically used terms, a VEI of 0 to 1 corresponds to a "Hawaiian" type eruption. Values of 1 to 3 are usually "Strombolian," 2 through 5 are "Vulcanian," 4 to 6 are "Plinian," and anything above that is "ultra-Plinian." These terms describe the explosivity of individual volcanic regions in comparison with the quiet production of lava. The volcanic activity in Hawaii is only rarely explosive. Stromboli and Vulcano, islands off the coast of southern Italy, are more explosive. The term "Plinian" refers to the eruption of Vesuvius in A.D. 79, which was described by Pliny The Younger. By European standards, this was a violent eruption. The VEIs for several well-known eruptions are:

TEPHRA VOLCANO	YEAR	VOLUME (m^3)	VEI
Nevado del Ruiz (Colombia)	1985	10^7	3
Galunggung (Indonesia)	1982	10^8	4
Mount St. Helens	1980	10^9	5
Krakatau (Indonesia)	1883	10^{10}	6
Tambora (Indonesia)	1815	10^{11}	7
Aegean (Mediterranean)	10,000 years ago	10^{12} (?)	8

Long before the introduction of the concept of plate tectonics, scientists recognized that volcanic provinces can be distinguished due to the characteristic behavior of the VEI in each province. The family of "Atlantic volcanoes" generally has lower VEIs than that of "Pacific volcanoes." The label "Atlantic" in the old nomenclature corresponds to the volcanism found along mid-ocean ridges, while "Pacific" describes volcanism at subduction zones.

The degree of explosivity of a volcano is especially influenced by two factors: the viscosity of the magma and the percentage of volatile components. The volatile components may be dissolved gases or gas bubbles. Water generally plays a dominant role among the gases. As the magma's viscosity increases, it becomes more resistant to deformation from the action of forces and mechanical stresses. This can lead to increased gas pressure because of the increased ability of the magma to withstand pressure without allowing gases to escape.

The magma rising at subduction zones is rich in silicates and water. The water content comes mainly from the sediments dragged along with the subducting lithospheric plate. At a given temperature, the viscosity of the magma decreases with increasing amounts of dissolved water. The effect is especially strong for silica-rich magmas of subduction zones where the

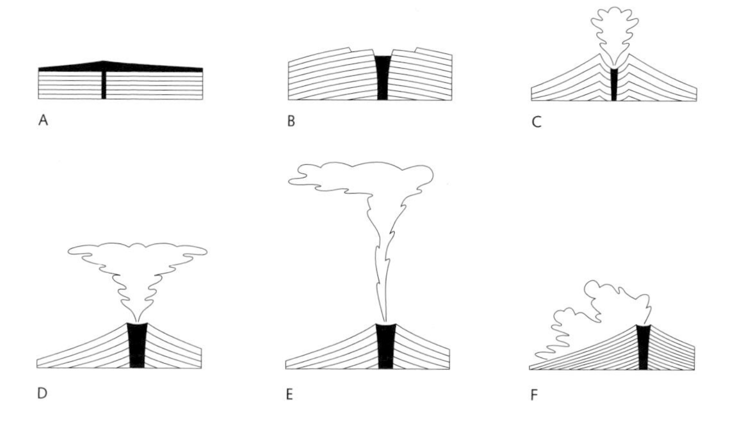

Nomenclature for volcanic activity.

A. Icelandic: produces large volumes of low-viscosity, basaltic lava that cover large areas (typical for Iceland).

B. Hawaiian: similar to Icelandic; however, after the lava flows out, large collapse structures called *calderas* develop (typical for the Hawaiian Islands).

C. Strombolian: usually short episodes of lava fountains separated by varying intervals (typical for the Stromboli volcano in the Tyrrhenian Sea, Italy).

D. Vulcanian: explosive activity that produces particles ranging in size from fine ash to volcanic bombs of several cubic meters volume. Episodes may last for several days (typical for the island Vulcano in the Tyrrhenian Sea, Italy).

E. Plinian: strong eruptions often reaching into the stratosphere, mainly with volcanic ash (named after Pliny The Younger who described the eruption of Vesuvius, Italy, in A.D. 79)

F. Peléean: usually occurs at the end of an episode of explosive activity, when the lack of thermal buoyancy leads to the collapse of an ash cloud and the descent of a *nuée ardente* (fiery cloud) along the slopes of the volcano (named after the eruption of the Mount Pelée volcano that destroyed the city of St. Pierre on the island of Martinique on May 8, 1902).

silicates combine to produce molecular chains and make the magma strong. As the magma rises, the lithostatic pressure decreases, and the solubility of water in the crystal structure is reduced. As the water is eliminated, the magma begins to crystallize more and its viscosity increases. In craters of subduction zone volcanoes, magma domes often develop over the conduit vent as plugs. Thanks to their strength, high gas pressure can develop under them. A particularly dangerous situation develops when the magma dome is not centered in the bowl of a crater, but is hanging on the side of a steep volcanic cone. If the dome slides because the slope is not stable, the magma underneath will foam because of the sudden drop in pressure. As a result, pressure waves and pyroclastic flows can develop. For example, a dangerous lava dome with an area of several square kilometers has developed during the past 15 years at the edge of the slope at the summit of Merapi volcano near the city of Yogyakarta in central Java, Indonesia.

The mafic magma at the mid-ocean ridges has far fewer silicates than subduction zone magma. Its viscositiey is therefore lower, and only low levels of pressure can develop.

The Eruption Products of Volcanoes

Magmatic products of volcanoes are classified by their phase when they leave the volcano. They are:

- volcanic gases
- solid, plastic, and melted ejecta (explosive activity)
- glowing, liquid, "coherent" flows of lava (effusive activity)

Volcanic Gases

The atmosphere of the Earth and the waters of the oceans are the results of volcanic degassing, which has been going on since tectonics started about three billion years ago. The exchange of material between the solid, the liquid and the gaseous Earth—between the lithosphere, the hydrosphere, and the atmosphere—is still going on. The gases, aerosols, and dust-sized particles emitted by violent eruptions can penetrate far into the stratosphere. At altitudes of 25 to 40 kilometers, they form clouds that spread throughout the world and influence climate everywhere. It is important to know the composition and amount of gases emitted by volcanoes, particularly in view of the increasing effects of human activities on the atmosphere. We can deduce the type, temperature, and pressure of the source magma from the chemistry of volcanic gases. Because gas and steam can rise rapidly from fissures extending deep into a volcano, they can also be an important precursor of eruptive activity.

Collecting gas samples from an eruption column is one of the most difficult and dangerous tasks of volcanology. In 1993, six volcanologists died on Galeras volcano in Colombia. They were surprised by an eruption while collecting gas samples. Because of the danger, remote techniques for studying gas emissions are preferred, despite the greater uncertainties. Infrared spectrometry is one of the most important methods. It is used to make a quantitative estimate of the amount of water injected into the stratosphere. The concentration of sulfur dioxide is measured with the help of a *co*rrelation *spec*trometer, or *cospec*. It measures how much ultraviolet light is absorbed by

The eruption of Vesuvius in 1779. Gouache by Pietro Fabris. The bulbous structure of the eruption cloud clearly shows the rhythmic ejection of volatile eruption products. It is reminiscent of steam coming from a steam engine.

(From G. B. Alfano and I. Friedländer, Die Geschichte des Vesuv. *Berlin, Germany: Verlag Reimer, 1929.)*

sulfur dioxide. Recently, attempts have been made to learn about volatile emissions of past eruptions by investigating fluid inclusions in solid volcanic products. This should help to estimate the volumes and products of giant eruptions in the historical and geological past.

Most geochemists now agree that water, H_2O, makes up 30 to 90 percent of the volatiles emitted by volcanoes and is therefore the dominant component. To give an example, between 1985 and 1990, about 1 cubic kilometer of water was released from magma during eruptive activity of Nevado del Ruiz, a volcano in Colombia. The second-most common component of volcanic gases is carbon dioxide, CO_2, usually 10 to 50 percent, followed by sulfur dioxide, SO_2, with 5 to 30 percent. The ranges show the great variation in the composition of gases. Other gases only make up a few percent of the emissions. Some of these trace gases are hydrogen (H_2), carbon monoxide (CO), carbonyl sulfide (COS), hydrochloric acid (HCl), hydrogen sulfide (H_2S), hydrofluoric acid (HF), methane (CH_4), mercury (Hg), and some noble gases. It is noteworthy that, aside from the overabundance of sulfur, the relative amounts of the gases are very similar to what we find in the atmosphere, the hydrosphere, and the sediments of the Earth.

Ejecta

The gas bubbles in rising magma increase in volume as the surrounding pressure decreases. How big the bubbles become depends on the viscosity of the magma. As magma degasses, its viscosity increases. The magma becomes stiffer and resists an increase in the size of the bubbles. Finally, the bubbles reach a maximum possible volume. As the bubbly flow reaches the surface in the volcano's crater, the pressure of the overlying magma decreases until the bubbles burst at atmospheric pressure. The pressure on lower layers drops, and the point at which bubbles develop moves slowly downward until it reaches a dynamic equilibrium with the new incoming magma. These processes are determined by many factors: thermodynamic parameters, viscosity and fracture behavior of the magma, the amount of volatiles and the velocity of new incoming material, among others. Because the processes depend on so many factors, there are many different types of eruption behavior. When the magma has a low viscosity, the degassing is usually quiet. Bubble-rich magma flows out as a continuous mass, a process that is called *effusion.* With highly explosive eruptions, a broad spectrum of solid products results, ranging from lava droplets that have cooled to ash, to scraps of still glowing, liquid lava that can land at great distances from the crater. In addition, there are blocks and rock splinters that have been torn from the sides of the conduit or crater. Pyroclastic flows and glowing avalanches (*nuée ardents*) are especially dangerous in the neighborhood of a volcano. They develop when an eruption column collapses because its thermal buoyancy is depleted. As a result, enormous amounts of loose material are deposited around the volcano in a short time.

Effusive Activity

At many volcanoes, the activity can be both explosive and effusive. Effusive activity is a non-explosive outpouring of lava. We often observe a situation where magma high in the volcano degasses, accompanied by violent explosions, while lower on the volcano's slopes degassed lava flows out quietly, almost like water, from a fissure, or *bocca* (the Italian word for mouth). This kind of lateral activity is typical for Mount Etna in Sicily and also for the volcanoes on Big Island, Hawaii. Remaining gas may still escape from the flowing lava with small explosions. The cone-shaped chimneys that develop, *hornitos*, are usually only a few meters high.

The shapes that develop when lava solidifies depend mainly on its viscosity, the remaining gas content, and especially the slope of the volcano. In masses of lava, temperature decreases and viscosity increases near the surface. The temperature at which lava solidifies is between 500 °C and 900 °C, depending on its chemical composition and gas content. If the lava is hot and has a low viscosity, it quickly develops a high viscosity skin, which is carried along if the lava is flowing. If the flow velocity is low, the large areas of skin remain smooth and even. We call this *pahoehoe lava*. "Pahoehoe" is a Hawaiian exclamation that expresses that the solidified flow is easy to walk on. In the transition from steep to flatter slopes, the flow velocity decreases. This can result in a dam when cooled lava plates on the surface that have been carried along by the flow pile up. One spectacular form of pahoehoe lava, *rope lava*, is often photographed. This structure is formed when the skin of the lava is still soft. In slow motion, the skin is rolled up by the lava flowing under it, creating "ropes." Because the middle of the flow is more rapid

Schematic representation of the arrangement of volcanic eruption products that are typical for flank eruptions of volcanoes such as Etna, Sicily. The drawing was made at the beginning of the twentieth century by August Sieberg, who was a professor of geophysics in Strasbourg and Jena, Germany, for many years. For the first time, it provided a view into the inner workings of a stratovolcano, a view that is still accepted today. (In contrast to the sketch, the conduits for magma are in reality only a few meters wide.)

(From August Sieberg, Einführung in die Erdbeben- und Vulkankunde Süditaliens. *Jena, Germany: Verlag G. Fischer, 1914.)*

than the edges, the ropes are usually bent and pushed together in a pile.

The surface of high viscosity lava is torn into blocks and scoria. This type of lava has received the name *a'a lava.* "A'a" is the Hawaiian cry of pain elicited when crossing the solidified lava barefoot. Very viscous lava rarely flows away from its point of exit from the volcano. Usually, it piles up to form a *lava dome.* If a lava dome develops over a conduit, it is called an *intrusive dome.* Gas can build up to high pressure under such domes. When the dome is suddenly blown away, the magma underneath foams up due to the sudden depressurization, and produces a *pyroclastic flow.*

If low-viscosity lava enters water, then *pillow lava* develops. When it comes in contact with water, the lava forms a thin, solid skin. The lava that follows breaks the skin and the process repeats itself. The result is a cauliflower-shaped lava formation with round, pillow-like bulbs of various sizes.

Volcanoes in the Cascades of North America

The San Andreas Fault enters the Pacific Ocean and joins with the Mendocino fracture zone in northern California. This zone extends far into the Pacific plate. Several hundred kilometers west of the North American coast, a wide, submarine mountain range branches off the Mendocino fracture zone toward the north. In the center of the range, there is recent volcanism with lava flows and many hot springs. This range is an oceanic ridge—that is, a divergent plate boundary. The most important segment of the ridge is the Juan de Fuca Ridge. It separates the Pacific plate from the Juan de Fuca microplate, which is being subducted under the North American plate.

The most visible sign of its subduction can be found about 150 kilometers east of the coast and parallel to the ridge: the volcanoes of the Cascade Range. From north to south, the volcanoes that have been active in geologically recent times are Mount Baker, Glacier Peak, Mount Rainier, Mount Adams, and Mount St. Helens in Washington; Mount Hood, Mount Jefferson, Three Sisters, Newberry Crater, and Crater Lake in Oregon; and Medicine Lake, Mount Shasta, and Lassen Peak in northern California. Most of these mountains have the typical shape of *stratovolcanoes*, a cone towering over the neighboring countryside. At the south end of the chain, the two most prominent peaks are Mount Shasta, 4300 meters high, and Lassen Peak, 3190 meters high. Mount Shasta is made up of several intermingled lava domes, ash cones, lava flows, and solidified pyroclastic material. It was most recently active in 1786. Large eruptions apparently occur every 600 to 800 years. Lassen Peak is made up of an astonishing number of lava domes. The most recent period of eruptive activity occurred from 1914 to 1917. Glowing avalanches and mudflows are the most likely hazards from these stratovolcanoes. When they are active, the heat can quickly melt the ice and snow covering them, producing huge amounts of water in a short time. This can lead to disastrous floods that reach far into the countryside.

About 7000 years ago, a giant eruption produced Crater Lake, a *caldera* in southern Oregon. A caldera is a depression formed by the explosion or collapse of a volcanic cone. During the eruption, about 50 cubic kilometers of magma were ejected as the upper 1000 meters of the stratovolcano Mount Mazama were blown away. The stump of the cone is the wall of the present-day caldera. It has a diameter of 10 kilometers and is 600 meters deep. Continuing activity after the giant eruption produced Wizard Island in the crater lake.

Mount Rainier, a steep, glacier-covered cone 4392 meters high, is visible from Seattle and Tacoma on clear days and poses a threat that should not be underestimated. Thick layers of loosely compacted volcanic deposits cover the flanks of the mountain. One of the biggest known mud- and ashflows, with a volume of more than 1 billion cubic meters and more than 60 kilometers long, was deposited during a phase when Mount Rainier's activity consisted of steam explosions. Cities and communities are now built on top of the deposits. Mount Rainier was most recently active in 1840. Today, there are only shallow earthquakes within the mountain and fumaroles and steam emission at the summit.

Because of its explosion on May 18, 1980, Mount St. Helens is one of the best known volcanoes in the Cascades. Geologically, it is very young. The oldest known deposits are about 50,000 years old. In the past 500 years, there have been four eruptions that were similar in size to, or larger than, the explosive activity of 1980. The recent phase of activity was therefore not a surprise to volcanologists. The volcano had, however, been quiet since 1857.

In March 1980, earthquake swarms occurred under Mount St. Helens. A short time later, small steam explosions and ash emissions from a new crater were observed on the summit. In April and early May 1980, this activity decreased. However, the entire north flank of the volcano began to bulge. The bulge grew about 1 meter per day until it stood out from the original cone by more than 100 meters. It was fairly clear that the bulge was being caused by rising magma. Without any warning, on May 18 an earthquake of magnitude 5.1 occurred under the mountain. Within a few seconds, the northern flank of the mountain broke off, causing a gigantic slide of ice, rock and mud. The sudden depressurization released the water-saturated, overheated magma in the mountain and allowed it to foam up. A pressure wave of hot gases, steam, ash, and debris propagated laterally away from the volcano at speeds up to 1000 kilometers per hour. The surrounding forests were destroyed. The pressure wave was recorded at earthquake stations around the world. At the same time, a Plinian cloud developed that reached an altitude of 20 kilometers. The eruption lasted about 9 hours. When it was finished, the summit of Mount St. Helens was missing. It was replaced by a crater 1.5 kilometers in diameter and 1 kilometer deep.

The strong volcanic activity in the Cascade Range, associated with comparatively weak seismic activity in the Juan de Fuca subduction zone, is a puzzle for many geophysicists. Some seismologists fear that deformation energy could be collecting in an unknown seismic gap in preparation for a strong earthquake.

Volcanic Eruptions and Climate

What does the writing of *Frankenstein*, one of the most famous horror stories of world literature, have in common with the French revolution? This may look like a trick question, but it has a deeper background. Among the causes of both events, volcanic eruptions were important factors. The eruption of the Tambora volcano on an island east of Java, Indonesia, in April 1815 played a role in the first event. Eruptions of the Laki fissure zone in Iceland in 1783 were involved in the second. Both eruptions were Plinian in strength and delivered huge volumes of gas and particles high into the atmosphere. The eruption of Tambora released 25 cubic kilometers of tephra that reached a maximum altitude of 45 kilometers. It also emitted more than 50 million metric tons of sulfuric acid (H_2SO_4), 200 million metric tons of hydrochloric acid (HCl), and 100 million metric tons of hydrofluoric acid (HF). The gases and particles from the two eruptions absorbed and scattered radiation from the Sun. As a consequence, the temperature near the ground decreased. The cold summers in the years following 1783 and 1815 caused poor harvests and starvation. In France, they led to the revolution in 1789. Cold storms and rain in the summer of 1816 kept Mary Shelley inside. She was on vacation at Lake Geneva in Switzerland, and in order to pass the time, she wrote the story of *Frankenstein.*[3]

Many studies are now investigating the influence of volcanic eruptions on the climate of the Earth. One of their goals is to allow us to distinguish between the natural emissions of the Earth and those of man, particularly carbon dioxide and chlorofluorocarbons. We would like to know whether volcanic eruptions increase the greenhouse effect and whether they damage the ozone layer.

One important problem in these studies is that the temperature of the atmosphere is controlled by many different mechanisms. Volcanism is only one factor. From observations of many large volcanic eruptions and the weather patterns in the years following them, there appear to be statistical correlations for certain types of changes in the weather. However, even these may not be completely reliable. For example, the eruption of Tambora occurred at a time when the temperatures in Europe were already below average. It is possible that Tambora was only the straw that broke the camel's back.

We can now collect instructive information about the chemical and physical condition of the high atmosphere using LIDAR measurements. LIDAR is the abbreviation for *Light Detection and Ranging.* Measuring stations on the Earth continually send laser impulses high into the atmosphere and observe the travel time and intensity of the back-scattered light. Several days after strongly explosive activity began at El Chichón in Mexico at the end of March 1982, LIDAR instruments on Hawaii measured especially strong back-scattering. It was associated with a dust cloud at an altitude of 25 kilometers. Stratospheric winds blow from the east toward Hawaii, and the arrival time of the cloud is consistent with the mean wind velocity of 70 kilometers per hour. The cloud continued to drift westward and in May it was measured with a LIDAR system in Germany. At that time, it had an altitude of 16 kilometers. Several months later, a second cloud was measured there at an altitude of about 25 kilometers. The two clouds slowly mixed with the atmosphere until they formed a single layer, which could be observed until the end of 1983. The colorful sunsets of 1982 and 1983 were a consequence of the scattering of light in the stratospheric dust layer.

Together with samples collected by balloons and airplanes, model calculations of the formation of clouds have changed the picture we have of the climatological effects of Plinian eruptions on several essential points. In contrast to our earlier assumptions, the climatological effects of volcano-caused atmospheric pollution do not depend primarily on the volume of the ejected ash and dust particles. Despite comparatively weak explosive activity, El Chichón produced haze with comparatively strong shielding characteristics. Detailed analyses have shown that sulfur dioxide is important in determining the density and scattering characteristics of a cloud. In a complex process and under the influence of sunlight and water vapor, sulfur dioxide is oxidized to gaseous sulfuric acid. The acid collects on dust particles and ions, and forms aerosols that scatter light. This photochemical process is somewhat delayed, however. Months can pass before the gas is completely converted into aerosols. The aerosols are much less dense than silicate particles from the volcano, so they remain in the air much longer. In addition, the small vertical temperature gradient in the lower stratosphere, from 10 to 30 kilometers, prevents extensive atmospheric mixing. The aerosols can remain suspended in the atmosphere for several years, which means that the effect of different eruptions that are separated in time can be cumulative. A few small eruptions can be just as important for the climate as one big one. How a volcanic cloud spreads through the atmosphere horizontally depends on seasonal weather patterns in the stratosphere.

The eruption of El Chichón carried around 3 to 4 million metric tons of sulfur dioxide into the stratosphere. The average temperature decrease in the affected zones was about 0.5 °C. As a comparison, the eruption of Krakatau in Indonesia in 1883 caused the average temperature in the eastern U.S. to decrease by about 1.5 °C. Climate is not only determined by the temperature. An increase in the amount of aerosols in the lower atmosphere (up to 10 kilometers) can also lead to an increase in the number of clouds and the amount of precipitation.

From 1960 to 1990, the worldwide volcanic production of sulfur dioxide is estimated to have been about 15 million metric tons per year. Only a small percentage of this is due to strong eruptions that deliver their emissions into the stratosphere. The rest come from *solfataras* and *fumaroles*. The activity of solfataras and fumaroles is characterized by quiet, continuous emission of gas and steam from cracks and fissures in the volcano. For example, between 50 and 150 metric tons of sulfur-containing gases are emitted every day by the Great Crater fumarole field on the island of Vulcano, Italy. This quiet emission is comparable to the exhaust of a factory chimney. The volcanic gases remain in the troposphere, the lowest part of the atmosphere. They cause at most 10 percent of the acid rain that falls. Mankind's contribution to atmospheric sulfur is an order of magnitude larger.

Do volcanic eruptions influence the ozone layer? In the winters following the gigantic eruption of Pinatubo in June

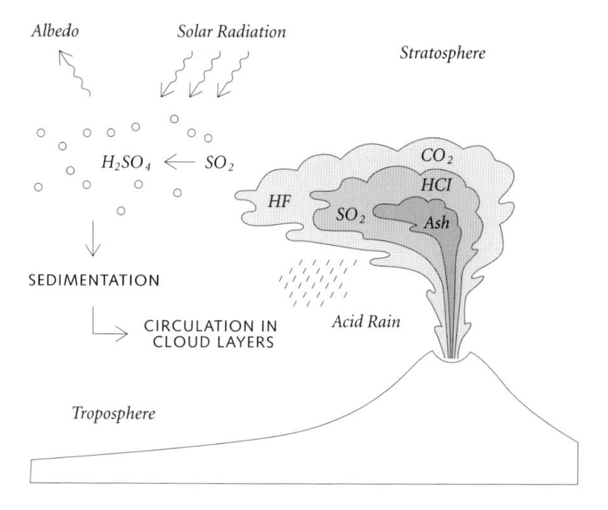

Effects of volcanic emissions in the troposphere and stratosphere.

(After a drawing by Peter Schick.)

1991, the ozone hole was unusually large. The effect was particularly noticeable at the higher latitudes of the Northern Hemisphere. Initial attempts to provide an explanation were contradictory. The aerosol concentrations in the stratosphere were 100 times higher than normal due to the eruption. According to calculations, they should have caused an increase, not a decrease, in the ozone concentration. A solution was finally found when the model calculations included the effects of the chlorofluorocarbons in the stratosphere. Evidently, the behavior of the ozone layer under the influence of volcanic gases is highly sensitive to the concentration of chlorine compounds. The addition of volcanic gases to the high levels of man-made chlorine compounds causes a decrease in the ozone concentration and therefore an increase in the size of the ozone hole.

Volcanoes That Should Not Exist

The islands of Hawaii are one of the best known and most active volcanic regions in the world. On the easternmost island, called Hawaii or the Big Island, lava has been flowing from a side crater in the Kilauea rift system continuously since 1983. It flows into the ocean and is a spectacular sight, particularly at night. Glancing at the map, you might be surprised. Hawaii is in the middle of the Pacific plate and lies thousands of kilometers from the nearest active plate boundary. According to the otherwise successful rules of plate tectonics, there should be neither earthquakes nor volcanism there. Hawaii is the best-known, but far from the only exception to the plate tectonic theory of volcanic activity. The volcanism on the Canary Islands, the Tibesti Massif in the Sahara Desert, and Reunion Island in the Indian Ocean cannot be explained by plate tectonics.

At the end of the nineteenth century, some geologists noticed that the volcanism in the chain of Hawaiian Islands gets older as one moves toward the northwest. Proceeding from Hawaii to Maui, Molakai, Oahu, and Kauai, the age of the islands increases. Northwest of Kauai, the westernmost of the Hawaiian Islands, the erosion of the volcanic islands is so far advanced that they can only be seen as high points on a map of the topography of the ocean floor. In 1960, the Canadian earth scientist Tuzo Wilson gave a plausible explanation for this observation. A stream of heat and material rises at the Hawaiian Islands from a source of heat, a *hot spot*, fixed in the lower mantle. Such streams are usually called *plumes* from the French word for feather. When the heat and material reach the lithospheric plate, melting occurs with the resulting development of volcanoes. However, the Pacific Plate is moving, gliding slowly over the hot spot. After a time, each newly created volcano moves away from the conduit of hot material that is anchored in the Earth. Slowly its volcanic activity dies. The process is like that of a factory chimney that occasionally gives off a puff of smoke. Wind passing the chimney carries the puffs away. From their track, we can derive the speed and direction of the wind. The track of the volcanic mountains constructed over the hot spot can be followed back over the course of 75 million years from the island of Hawaii to the vicinity of Kamchatka. It agrees perfectly with the direction and speed determined for the Pacific plate from other observations.

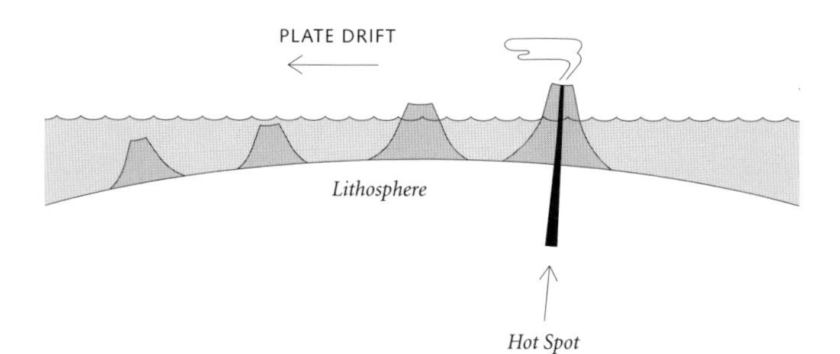

Development of volcanic chains on a moving lithospheric plate due to hot spot conduits anchored in the Earth's mantle. The development of discrete volcanic cones can be explained by changes in the amount of magma produced as a function of time.

Today, we recognize more than one hundred volcanic areas associated with hot spots. Not all of them lie, like Hawaii, far from the active plate boundaries. It is likely that Iceland, the Azores, and Tristan da Cunha, all volcanic islands on the Mid-Atlantic Ridge, are fed by hot spots. At each of these places, a much larger volume of magma is produced than is normal for the relatively small, submarine volcanoes along the ridge.

All investigations of the positions of hot spots indicate that they are long lasting and relatively fixed in the body of the Earth. This is only possible if the sources of the hot spots lie below the depth at which convection currents move horizontally. Thus, they cannot be anchored in the upper mantle. Only the high viscosity of the material in the lower mantle can make for a stable location over long periods of time. Although we do not yet have any direct indications, many researchers believe that the source of the hot spots can be found at the core-mantle boundary.

Despite the speculation about the source and development of hot spots, they make an important contribution to the description of global tectonics. Because they last a long time and are fixed in the Earth, they provide an excellent reference system for the moving lithospheric plates. In plate tectonics, we assume that the African plate has been stationary for the past 30 million years. The hot spots on the African continent support this assumption. At each of the volcanic regions—Tibesti, Mount Camaroon, Nyiragongo, and Reunion Island—many millions of years worth of lava are piled on top of each other. This is evidence of the fixed position of the African plate. The opening of the Atlantic Ocean between South America and

Africa was not caused by symmetrical movement of both continents relative to the Mid-Atlantic Ridge, as we might suppose. In an absolute reference system, Africa's position has remained fixed relative to the Earth's mantle, and the Americas have moved away from it toward the west at twice the speed of the Mid-Atlantic Ridge.

It is likely that hot spots are also important in initiating the phases of great tectonic change. In the 1930s, the German geologist Hans Cloos described uplifted regions in the structure of continents. From their centers, three fracture zones radiate like the arms of a Mercedes star. The uplifting probably occurs when a continent comes to rest over a hot spot. Two of the three fracture arms can open to form a new ocean. The third arm runs into the continental landmass without giving rise to an ocean. When we reconstruct Gondwanaland, the continent that broke up 120 million years ago to make South America and Africa, we can recognize many such structures. There are clues that the breakup was associated with weak zones in the lithosphere resulting from hot spots. A similar process is presently occurring where the Arabian Peninsula is separating from the African continent. The Red Sea and the Gulf of Aden are the arms opening to form ocean. The third, "dry" arm is the Afar Triangle and the Ethiopian rift system.

Volcanism in the Solar System

Earth is not the only planetary body in our Solar System with volcanism. Nonetheless, it remains a unique planet with regard to volcanic phenomena. With the possible exception of Venus, convection cells and the resulting phenomenon of plate tectonics could develop only in the mantle of our planet. This is because the thermal radiation from a body depends on the ratio of its surface area to its total volume. The bigger the ratio, the more heat will be lost per unit time. The amount of heat that developed in a heavenly body during its formation due to the dissipation of kinetic energy from the impacts of planetesimals can be found from the ratio of the size of the metallic core to the rocky mantle. If we set the core-to-mantle ratio as 1 for Earth, then Venus has a ratio of 0.9, Mars has 0.8, and the Moon has 0.1. If we disregard Mercury, Earth collected the most heat during its origin and, at the same time, it loses heat most slowly. Thus, of all the terrestrial planets, it is the least "burned out."

Let us consider now the volcanic development of the Moon and the inner planets.

Our Moon

The great plains on the Moon have few craters and were originally thought to be oceans of water, and are therefore called *maria* (Latin for seas). They are flood basalts laid down 3 to 4 billion years ago. Chemically, they are very similar to terrestrial basalts. However, they have somewhat more iron and magnesium, and with fewer alkalis and less water. From the dimensions of the maria, we can conclude that the outflow volume was on the order of several cubic kilometers per day. On the Earth, only the flood basalts of the Deccan Traps plateau are similar in size. Very few of the volcanic cones that are typical for the Earth are found on the Moon. This is mainly because of the low viscosity of the Moon's lavas. The low water content of the lavas means that the Moon volcanoes were not very explosive, so that no stratovolcanoes developed.

Mars

On Mars, like on the Moon, we find extended lava fields. The flows can stretch for hundreds of kilometers, even though the slopes are very gentle. This is a consequence of the high temperature and low viscosity of the magmas. For unknown reasons, the volcanism is concentrated in the Northern Hemisphere. There we find the largest shield volcanoes known in the Solar System. Shield volcanoes are relatively flat and get their name by resembling a shield. The biggest of them, Mons Olympus, is 25 kilometers high. It has a ring wall 300 kilometers away. The structure of the shield volcanoes on Mars is comparable to that of the great shield volcanoes Mauna Loa and Mauna Kea on Hawaii. There are collapse craters in the summit region, large linear lava flows on the flanks, and extended channels that probably represent collapsed lava tubes. The difference in the sizes is plausible. The volcanoes of Hawaii wander with the lithospheric plates over a hot spot fixed in the Earth's mantle. The volcanic products are therefore distributed throughout the entire chain of volcanoes. In the stable Martian crust, however, the volcano remains stationary over the source of magma. The mountain piles up until the increasing pressure in the magma column prevents new magma from rising.

Shield volcanoes are not the only volcanic structures on Mars. Alba Patera is a relatively flat lava-covered region more than 1600 square kilometers in area. The lower-lying plains are often covered by thick ash and sediment layers. This is an indication that erosion through wind, glaciers, and probably water has played an important role in Mars's history.

The age of the Mars volcanoes is subject to controversy. The age of a surface structure is usually determined by the number of meteorite craters that are found in the area. More impact craters means an older structure. Recent volcanism smoothes out old craters. In order to determine the age of the volcanic activity on Mars, we operate on the assumption that the rate of impact of meteorites on Mars is the same as that on the Moon. However, we cannot prove this assumption. For the Moon, the density of impact craters was compared with the absolute age of rock samples taken from them. Using this data, some researchers find that the volcanic activity at Mons Olympus goes back 200 million years. If this number is correct and if Mars is four billion years old, then we conclude that volcanism may still be active on Mars.

Venus

Because its mass, density, and diameter are similar to those of Earth, Venus is often called the Earth's twin. Thus, it would be understandable if there were also similarities in the tectonic processes affecting both planets. It is, however, difficult to investigate Venus's surface directly because of its dense atmosphere. It was not until the arrival of the American space probe Magellan that the topography of Venus was mapped, with a precision of 300 meters by using radar. The surface of Venus is covered with well-preserved, ring-shaped impact craters. They are estimated to be less than a few hundred million years old. Older impact craters were apparently covered by an extensive phase of volcanic activity.

The surface of the planet has other structures that must be related to tectonic activity. There are zones of compression and extension; linear, extended fractures that are possibly faults; folded mountains; and shield volcanoes over areas of large, dome-shaped uplift. In contrast to the Earth, where tectonic activity is concentrated at plate boundaries, the tectonically active regions appear to be distributed evenly over Venus's surface.

The Moons of Jupiter

The driving forces and products of volcanism on the inner planets are, in principle, the same as those we find on Earth. For the outer planets and their moons, they are completely different. The volcanism on Io, the innermost of Jupiter's large moons, is a good example. Using images of Io from the space probes of the Voyager series, we discovered the most active volcanoes in the Solar System. Material is thrown geyser-like on ballistic trajectories hundreds of kilometers above the eruption vents. Many hot calderas with lava flows can be found on the surface. The entire surface of Io is free of impact craters, a sign of the high rate of production of volcanic material.

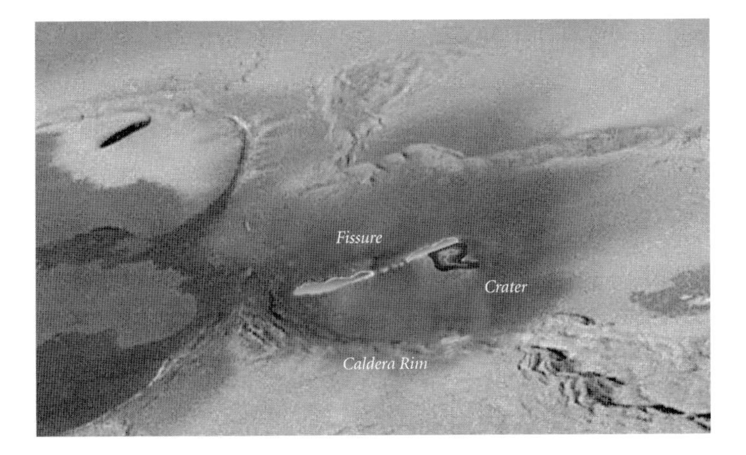

Image of volcanic structures and volcanic activity on Jupiter's moon, Io, taken by NASA's Galileo spacecraft on November 25, 1999. Sulfur dioxide at a temperature of more than 1200 °C is emitted from a 25-kilometer-long fissure (center). When it reaches the vacuum at the surface, the liquid sulfur dioxide evaporates to become a gas. Then the gas freezes and is deposited as "frost" on the ground. The image clearly shows the steep walls of an eruption fissure surrounding a big caldera. Calderas on Io can be several hundred kilometers in diameter. They are the largest-known collapse structures in the entire Solar System.

(NASA and the California Institute of Technology, Jet Propulsion Laboratory, Pasadena, California.)

Spectral analysis shows the dominance of sulfur in Io's emission products. We cannot completely exclude the possibility of silicate-based volcanism, however. The topography on Io changes by several thousand meters, which cannot be explained by a crust that includes large deposits of hot, soft sulfur. We assume that the heating of Io is caused by friction due to the tidal forces of Jupiter. Io orbits Jupiter at varying distances on a path that is disturbed by the other Galilean moons. The changes this causes in the gravitational forces gives mechanical energy to Io, which is dissipated into heat. Another source of energy could be an interaction with the magnetosphere of Jupiter. This could induce an electric current in Io, which would produce heat.

Europa, the next moon out from Jupiter, has volcanism of a completely different type. Again, the lack of impact craters indicates the continual renewal of the surface through tectonic or volcanic processes. The mean density of Europa implies that it has a core of silicate rocks that is covered by a thick layer of ice. The ice layer is cut by a pattern of fractures from which steam escapes. As with Io, the cause is assumed to be heating as the result of tidal friction. On Europa, however, it is at least a factor of 10 weaker than on Io.

Callisto is Jupiter's largest moon. Its surface is covered with meteor craters. If volcanism occurred during its history, it must have ceased long ago.

Volcanic Hazard and Monitoring

In contrast to earthquakes, in which a static situation becomes dynamic within a few seconds and thereby instantaneously creates an event, the flow processes that drive volcanic activity take some time to develop before the activity changes. It should therefore be easier to estimate the short-term hazard associated with a volcano at any given time than it is to predict the strength of an earthquake and the time at which it will occur. However, the following case studies show that while it is possible in principle to protect a population from the dangers of volcanic activity, there are many social problems involved in practice.

After 1980, the activity level of a series of volcanoes, some of them well monitored, changed. Among them are the Phlegraean Fields near Naples, Italy, El Chichón in Mexico, and Izu-Oshima near Tokyo, Japan.

For Europeans, the activity at the Phlegraean Fields was the most exciting of these three. From 1982 until 1984, the city of Pozzuoli was subjected to a swarm of earthquakes that occurred very near the surface. The ground rose several millimeters per day, and after two years the uplift had reached its maximum, about two meters. From experience, many volcanologists concluded that these phenomena must be caused by the injection of magma into layers near the surface. Historically, the region is known for its strong volcanic activity. For example, within a few hours in 1538, a 150-meter-high volcanic cone now called Monte Nuovo was built. This implies that the volcanic hazard in this densely populated area is very high. Twenty-two thousand people were evacuated from the city of Pozzuoli. Still, despite all the precursors, the feared volcanic catastrophe did not occur. Today, no one can say whether these unusual occurrences may or may not be precursors to a volcanic catastrophe the next time they occur.

The situation in 1982 at El Chichón in Mexico was completely different. The previous eruption had taken place many thousands of years before and the volcano was considered extinct—when people even thought of it as a volcano. A seismograph station was installed only to monitor earthquake activity at a nearby reservoir. A few weak earthquakes that were recorded in March 1982 received little notice. At the end of March, the seismograph recorded signals that would have alarmed any geophysicist with experience in volcano seismology. The signals were not caused by impulsive, sporadically occurring tectonic earthquakes, but were regular oscillations. Volcanologists know such *volcanic tremor* well. The ground motion is caused by the noises of flowing magma near the surface. The next day, the mountain opened in a mighty eruption that killed several thousand people.

Could the population have been saved by scientists experienced in monitoring volcanoes? Probably not. The occurrence of flow noises indicates the presence of rapidly flowing, gas-rich magma near the surface. This increases our estimation of the likelihood of a dangerous eruption in the case of a volcano with a closed crater that has not been active for millennia. However, no one would have been able to organize the evacuation of an unprepared population within a few hours based on the suspicion of possible activity. The following example shows that evacuations are problematic, even in a country like Japan where the population is better prepared for earthquakes and volcanic eruptions than in any other place in the world.

In November 1986, strong explosive activity began on Izu-Oshima, a volcanic island in the Bay of Tokyo. The lava fountains reached heights of 1500 meters above the vent. Although villages several kilometers from the vent were not directly threatened, the population of the island was evacuated. From previous eruptions, scientists knew that devastating explosions were often associated with lava ejection because of the nearby ocean. Such steam explosions develop when the rising magma comes into contact with water bearing layers. The well-known maars of the Eifel region in Germany were probably created by such explosions. Despite a scientifically supported evaluation that a threat from the volcano to the island continued, the evacuation of 10,000 residents to the nearby mainland could not be maintained for more than two weeks. Under pressure from the evacuated population, the governor of Tokyo visited the island and declared, in opposition to the opinion of most of the Japanese volcanologists, that according to his "intuition," the volcano was quiet and the residents could return to their island.

The volcanic threat to life and limb is smaller than is usually assumed. In the past five hundred years, about a quarter of a million people have lost their lives due to volcanic activity, either directly or indirectly through famine caused by destruction of agriculture. This is a small number compared with the number of deaths from earthquakes. In the July 1976 earthquake in Tangshan, China, a similar number of people lost their lives. According to a saying in Central America, "You can run away from a volcanic eruption, but not from an earthquake." This attitude disregards one important point, however. Earthquakes have an upper limit in strength. Worldwide, an event the size of the Tangshan earthquake occurs on average once every century. In contrast, we do not really know the maximum level of destruction that can be associated with a volcanic eruption, just as we do not know how often such eruptions occur. We can only get a feeling for the possible size of eruptions by looking at the big calderas that we find on the Earth. *Calderas* are basically the remains of a volcano after it has ejected huge volumes of magma. The name was coined by the German Leopold von Buch after he visited the big Caldera di Taburiente on the island of La Palma in the Canary Islands. Toba Lake in northern Sumatra is one of the largest known calderas. It is almost 100 kilometers long. We estimate that 75,000 years ago, more than 1000 cubic kilometers of ash were ejected from it in the space of a few days. After the loss of material, the caldera collapsed to a depth of about 2000 meters. Later magma intrusions raised its floor again, so that today the island Samosir rises from the middle of the deep lake.

The devastating effects of such gigantic eruptions on man and his environment are probably only exceeded by those of a meteor impact. During the development of the Toba caldera, an area of 400 square kilometers collapsed and was then covered with ash. It effectively disappeared from the map. At the edge of the caldera, ash layers are several hundred meters thick. And at a distance of a few hundred kilometers, the ash cover is still several meters thick. Over an area twice the size of New Jersey, all human, animal, and plant life was buried and destroyed. We

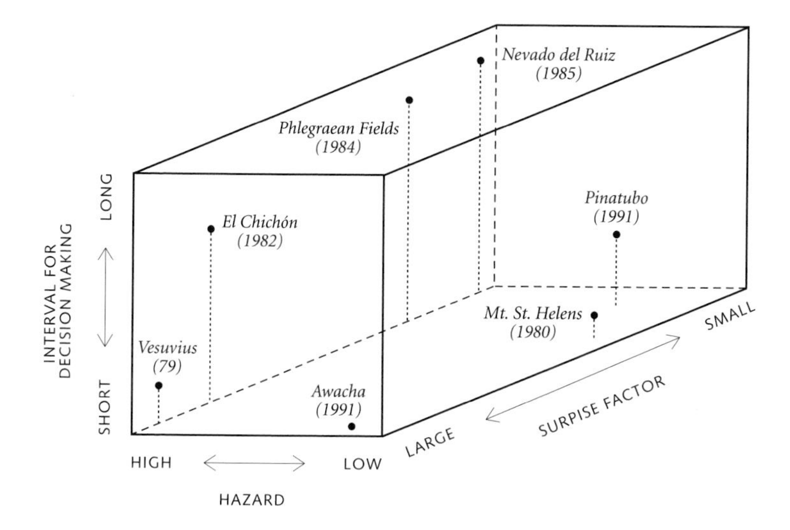

"Herrmann's Crisis Cube" classifies possibly catastrophic changes in the condition of a volcano according to their surprise factor, the level of danger, and the amount of time available for decision making. The figure shows several volcanic eruptions with the year of their occurrence. In risk evaluation, the values of the coordinates can be considered to be control parameters in the theory of nonlinear processes.

(After a drawing by Peter Schick.)

do not know how often such events occur. Gigantic eruptions like that at Toba Lake probably only happen every 500,000 years. Calderas with smaller diameters are, of course, more common. About every 50,000 years, collapses with diameters of 20 to 30 kilometers occur, and collapses with 5- to 10-kilometer diameters probably every 1000 years.

It is a basic goal of volcanology to investigate what causes the change from an inactive or weakly active state to an active phase, as well as the accompanying phenomena. We can only deduce from observations and plausible arguments that over the long term, increased activity is a consequence of the rise of a batch of magma from the depths. We cannot know why, when, where, or how big. In the short term, we must take factors into account such as those described on pages 98–106.

Our monitoring of volcanoes with the intention of estimating the future development of their activity is very similar to the description of weather prediction. The behavior of both systems is decided by types of flow. In contrast to the movement of air masses, the movement of magma is underground and cannot be observed directly from the Earth's surface. We are therefore limited to indirect measurements. The interpretation of the data is nearly always burdened with uncertain or even speculative assumptions.

With the help of modern measurement techniques, we attempt to detect the movement of new magma into the volcano or changes in a flow pattern that has been constant for years. We use the fact that magma differs significantly from its environment in is mechanical, thermodynamic, electrical, and chemical behavior. In simple terms, the transport of magma in the inside of the volcano can occur in two ways:

- Magma flows inside the volcano in preexisting, plate-like channels, or dikes. In many volcanoes, for example the walls of Somma at Vesuvius or in the Valle del Bove at Etna, we can easily recognize dike structures of cooled magma. Most of the vertically oriented dikes are about one meter thick. Vertically and horizontally, they can be several hundred meters long.
- The volcanic edifice is subject to mechanical loads, a consequence of its weight and the regional tectonic stress. New magma can enter through areas of weakness, preexisting cracks, or surfaces with minimal compression. It can then create a new conduit and a new vent.

Monitoring volcanoes is mainly based on ideas of magma transport. We can use seismographs to "listen" to the flow noises described on pages 100–101. From the characteristics and the tones of the noise and with the help of theoretical and experimental knowledge of flow acoustics, we can distinguish between different types of flow. We think, for example, that we can recognize whether the movement in the conduit is primarily unidirectional (monopole flow) or a convection current (dipole flow). The closed circuit of a warm water heater in a building is an example of a dipole flow. If water leaks from a hole in one of the radiators, then a monopole flow is superimposed on the dipole flow. The transition from dipole to monopole flow can be an indication that an eruption is coming.

Rapid and clearly defined fluctuations in the strength of recorded seismic signals are indications of gas bubbles in the flowing magma.

New conduits develop when cracks widen as the magma presses in. The widening is accomplished partly by plastic deformation, partly by fracturing. With seismic networks, we can observe the location and strength of the fracturing. In that way, we can follow the path of the incoming magma. At Kilauea, Hawaii, the upward flow of intruding magma was successfully tracked by seismographs—instruments that measure ground acceleration. They can be very sensitive, able to measure 10^{-12} times the gravitational acceleration. But even such sensitivity is not enough to find evidence of, and locate changes in, the flow of magma that takes place over years.

Very slow, asymmetric magma transport in the volcanic edifice, directed toward the Earth's surface or radially away from the central conduit, can lead to deformation of the edifice. Such deformation can be monitored using various methods. Traditionally, geodetic methods such as leveling and triangulation are used. Since these methods are very time-consuming and require considerable work, measurement procedures using satellite geodesy are preferred today. The *Global Positioning System* (GPS) allows us to make determinations of the locations of selected points with a precision of centimeters or less. Using Synthetic Aperture Radar (SAR) interferometry, volcanic regions can be sampled by satellite without the aid of ground stations. With repeat measurements taken during many orbits, deformation can be investigated. Since the beginning of the twentieth century, tiltmeters have often been used not only in earthquake regions but also on volcanoes for monitoring ground deformation. Tilt can be measured either with long, water-filled hoses in which changes in the level of the water are observed, or they can operate on the same principle as a crookedly hung door. Fifty years ago in Hawaii, these instruments were used to demonstrate that uplift on a volcano's surface was associated with the entrance of magma. Collapses, on the other hand, were associated with the outflow of magma.

The density of magma is different from that of the surrounding rock in the volcanic edifice. Movement of magma in the mountain will have an effect on the gravitational field of the volcano. This can be measured with high-precision gravimeters. However, various disturbing influences, such as changes in the ground water level, make the interpretation of such measurements difficult.

Intrusions of magma lead to increases in the SO_2 emissions. If no expensive gas monitoring instruments are available, lower pH values in crater lakes can be an indication of this effect.

In addition to the instruments and physical methods for monitoring volcanoes, qualitative observations are also used. Even when the volcano is extremely dangerous, these are often the only way to detect changes in its activity level. Unusual snow melt or glacier movement can be an indication that the temperature in the volcano has increased. Often landslides, increases in the number of rock falls, or the development of fissures are a result of the volcano becoming steeper due to rising magma.

Our knowledge about the factors leading to a volcanic eruption and their interactions are not nearly good enough to clearly follow the development of a volcano's activity from measurements. Mostly, we are limited to general estimates, plausibility arguments, and comparisons with previous eruptions of the same volcano. The most important question for the population affected by the volcano is whether the increasing activity will lead to a large and dangerous eruption. This still cannot be answered by volcanologists. Meteorologists have the same problem. The weather conditions that develop into rare weather phenomena like hurricanes or unusually strong hail can usually only be explained after the fact. The analogy with weather is not surprising. Volcanic and meteorological processes are controlled by flow patterns following nonlinear laws that depend on many parameters. When the system has reached a critical state, small changes can have large effects. Still, there are states in which it is possible to calculate what will happen with great certainty. A stable high-pressure system with little change in air pressure can be compared with the situation at Etna on Sicily. There the amplitude of the volcanic tremor only changes a little over time. But useful prediction should deal with change. However, it is difficult to tell whether a sudden falling of the barometer during a thunderstorm portends a weather catastrophe or only a storm that will break a few branches. It is just as difficult to use sudden changes in some measurement from a volcano to determine whether a change in activity will develop into a threat to nearby inhabitants.

1 Karl Sapper, *Vulkankunde* (Stuttgart, Germany: Verlag Engelhorns Nachf., 1927).
2 Sir William Hamilton, *Beobachtungen über den Vesuv, den Ätna und andere Vulkane* (Berlin, Germany: Haude und Spener, 1793).
3 Gordon Woo, *The Mathematics of Natural Catastrophes* (London: Imperial College Press, 1999), 5.

Further Reading and Information

Books

General Geophysics, the Structure of the Earth, Plate Tectonics

Brown, Geoff, Chris Hawkesworth, and Chris Wilson (eds.). *Understanding the Earth.* Cambridge, Massachusetts: Cambridge University Press, 1992.

A multidisciplinary summary of today's knowledge about the interior of the Earth and the dynamic processes occurring within it. The emphasis is on lithospheric processes. Intellectually challenging, but clearly written text with good illustrations.

Van Andel, Tjeerd H. *New Views on an Old Planet.* 2nd ed. Cambridge, Massachusetts: Cambridge University Press, 1994.

An exciting, in-depth introduction to the geological history of the development of continents, oceans, and the atmosphere.

The first edition won the 1986 Professional and Scholarly Publishing Physical Sciences Award of the Association of American Publishers.

Condie, Kent C. *Plate Tectonics and Crustal Evolution.* 4th ed. Woburn, Massachusetts: Butterworth-Heinemann, 1997.

A book that delves into the details of today's knowledge and understanding about the development of the Earth's crust and mantle since our planet's beginnings. It assumes a basic understanding of the earth sciences.

LeGrand, H. E. *Drifting Continents and Shifting Theories.* Cambridge, Massachusetts: Cambridge University Press, 1988.

The long path to plate tectonics. This book about the history of the development of our ideas about the dynamics of the Earth is worth reading. The narrative tells the stories of the many scientists who participated in the development of the theory and about their controversial ideas that were not always free of ideologies.

Earthquakes

Bolt, Bruce A. *Earthquakes and Geological Discovery.* New York: W. H. Freeman & Company, 1993.

Written by one of the grand old men of earthquake seismology, this book gives an overview of the causes and effects of earthquakes. It is suitable even for the uninitiated. The good pictures, the interesting examples selected from the pool of earthquakes that have occurred worldwide and an understandable text make the book worth reading.

Yeats, Robert S., Kerry Sieh, and Clarence R. Allen. *The Geology of Earthquakes.* New York: Oxford University Press, 1997.

A book for readers who are interested in a more detailed look at the consequences of earthquakes from the viewpoint of geotechnical and construction applications. Among other phenomena, it includes descriptions of liquifaction and landslides. Many case examples.

Collier, Michael. *A Land in Motion.* Berkeley, California: University of California Press, 1999.

A wonderful travel book for everyone who is interested in the San Andreas Fault in California.

Volcanoes

Decker, Robert, and Barbara Decker. *Volcanoes.* New York: W. H. Freeman & Company, 1997.

An introduction to all the aspects of volcanology, written by an experienced volcanologist and a science writer. Worthwhile and entertaining reading for both newcomers to volcanology and readers with background knowledge.

Fisher, Richard V., Grant Heiken, and Jeffrey B. Hulen. *Volcanoes—Crucibles of Change.* Princeton, New Jersey: Princeton University Press, 1998.

A richly and well-illustrated work, which deals especially with the effects of volcanism on mankind. The eruption of Mount Saint Helens is described in detail. The section "The Volcano Traveler" deserves particular mention. It is a travel guide to many volcanic areas in the world.

Decker, Robert, and Barbara Decker. *Volcanoes in America's National Parks.* Union Lake, Michigan: Odyssey Publications, 2001.

A guidebook recommended for all who wish to visit and hike at active and dormant volcanic areas in America's national parks.

Wood, Charles A., and Jürgen Kienle. *Volcanoes of North America: United States and Canada.* Cambridge, Massachusetts: Cambridge University Press, 1992.

Detailed descriptions of each of the volcanoes in the U.S. and Canada.

Frankel, Charles. *Volcanoes of the Solar System.* Cambridge, Massachusetts: Cambridge University Press, 1996.

A detailed and comparative look at volcanism on the planets and moons of our Solar System.

Magazines

Earthquakes & Volcanoes, published by the U.S. Geological Survey. For sale on CD-Rom by the Superintendent of Documents, U.S. Government Printing Office, Washington, DC, tel. 202.512.1800.

Earthquakes & Volcanoes provides information on earthquakes and seismology, volcanoes and related natural hazards of interest to both generalized and specialized readers.

Bulletin of the Seismological Society of America (B.S.S.A.), published by the Seismological Society of America, El Cerrito, California, ISSN 0037-1106.

A scientific journal covering seismological topics.

Bulletin of Volcanology. Journal of the International Association of Volcanology and Chemistry of the Earth's Interior (I.A.V.C.E.I.). Published by Springer-Verlag New York, ISSN 0258-8900.
A scientific journal covering volcanism-related topics.

World Wide Web

There are many excellent websites on the World Wide Web dedicated to topics related to earthquakes and volcanoes. Some include extensive catalogs of earthquakes and volcanic eruptions. There are many descriptions of individual volcanoes, often including travel and hiking information, for example for Etna and Stromboli in Italy. Current and up-to-date information is also available about earthquake and volcanic activity. It is possible to observe and follow eruptive activity via live cameras. Seismograms of earthquakes can be downloaded in near-real-time. The following sites are maintained by well-known institutions. In addition to extensive and detailed information, they also provide a good selection of links to other sources:

http://www.iris.edu/

IRIS is a university research consortium dedicated to exploring the Earth's interior. The home page includes an interactive educational display of global seismicity that allows the visitor to monitor earthquakes in near-real-time, view records of ground motion, and visit seismic stations around the world.

http://www.geology.sdsu.edu/how_volcanoes_work/

This web site is an educational resource that describes the science behind volcanoes and volcanic processes. The site is sponsored by NASA under the auspices of Project ALERT (Augmented Learning Environment and Renewable Teaching).

Index

Illustrations are indicated by page references in italics.

F

W

X